Mild Traumatic Brain Injury

Mild Traumatic Brain Injury

A Science and Engineering Perspective

Mark A. Mentzer

CRC Press
Taylor & Francis Group
Boca Raton London New York

CRC Press is an imprint of the
Taylor & Francis Group, an **Informa** business

First edition published 2021
by CRC Press
6000 Broken Sound Parkway NW, Suite 300, Boca Raton, FL 33487-2742

and by CRC Press
2 Park Square, Milton Park, Abingdon, Oxon, OX14 4RN

© 2021 Taylor & Francis Group, LLC

CRC Press is an imprint of Taylor & Francis Group, LLC

Library of Congress Cataloging-in-Publication Data
Names: Mentzer, Mark A., author.
Title: Mild traumatic brain injury : a science and engineering perspective /
Mark A. Mentzer.
Description: First edition. | Boca Raton : CRC Press, 2021. | Includes
bibliographical references and index. | Summary: "Mild traumatic brain
injury, also known as chronic traumatic encephalopathy, presents a
crisis in contact sports, the military, and elsewhere. This book reviews
current understanding of mTBI, methods of diagnosis, treatment, policy
concerns, and emerging technologies. This book details the
neurophysiology, epidemiology of brain injuries by presenting disease
models and descriptions of nucleating events"—Provided by publisher.
Identifiers: LCCN 2020023636 (print) | LCCN 2020023637 (ebook) |
ISBN 9780367362607 (hardback) | ISBN 9780429344947 (ebook)
Subjects: MESH: Chronic Traumatic Encephalopathy | Brain Concussion |
Biomedical Technology
Classification: LCC RC394.C7 (print) | LCC RC394.C7 (ebook) | NLM WL 354 |
DDC 617.4/81044—dc23
LC record available at https://lccn.loc.gov/2020023636
LC ebook record available at https://lccn.loc.gov/2020023637

ISBN: 978-0-367-36260-7 (hbk)
ISBN: 978-0-367-56776-7 (pbk)
ISBN: 978-0-429-34494-7 (ebk)

Typeset in Times
by codeMantra

Contents

Foreword

Consequent to my work at the US Army Research Laboratory (ARL) developing high-speed imaging, custom illumination devices, and cineradiographic X-ray systems, all in support of improvements in personnel protective equipment (PPE), I explored the biochemical mechanisms involved in brain injuries. My goal was to elucidate the fundamental processes in brain injury—especially mild traumatic brain injury (mTBI)—to better address the need for adequately instrumenting live-fire tests to assess mTBI in the most relevant manner. Despite the focus on the high prevalence of brain traumas experienced in the military and in contact sports, we lacked adequate means for instrumenting live-fire tests to assess mTBI. This limited the effective mitigation or prevention of mTBI through improved PPE or early diagnosis in the field coupled with appropriate therapeutic intervention.

I became interested in translational TBI research, which led to my development of a mTBI sensor to accurately correlate biomechanical data to biomedical and to identify thresholds for neuronal injuries. I abstracted from the literature the relationship between neuronal injuries and their insult thresholds and calibrated my sensor outputs to these injury modalities. This made it possible to identify soldiers in the field requiring treatment and to predict from specific injuries the most appropriate therapies. I then designed a field-portable diagnostic for biomarkers, which I did as a lab-on-a-chip microfluidic assay that detects protein biomarkers and guides further diagnosis.

I validated my sensor by designing a test chamber that uses a fluid-percussive injury apparatus and replicated injury-based biomarker medical data from the literature. I then correlated this data to sensor outputs, thus providing accurate assessment of injuries and such neurological damages as diffuse axonal injury and axon-glia dysfunction. This also makes it possible to relate injury thresholds to electrophysiological measurements of human function. This provided the basis for more accurate injury prediction and for relating my sensor data to models for tissue damage. The work culminated in Patent No. 9080984 included here as Appendix 1.

During this time, I published my third book (Mentzer, 2011) and engaged in graduate studies at Johns Hopkins University in biochemistry and molecular genetics. By attending a variety of imaging conferences, chairing brain injury conference sessions, and participating in Keystone and other neuroscience conferences, as well as numerous ARL committees, along with sponsoring academic initiatives with the Johns Hopkins Institute for Nanobiotechnology and other sponsored university research in embedded fiber impact strain sensors, I gained perspective in the field resulting in the motivation for this book. It has been my pleasure to participate in the exciting realm of neuroscience and to lend insights into the issues of mTBI.

REFERENCE

Mentzer, M.A. 2011. *Applied optics, fundamentals and device applications-nano, MOEMS, and biotechnology.* New York: CRC Press Taylor and Francis Group.

Acknowledgments

I would like to thank my friends and colleagues—scientists, engineers, and medical personnel—for numerous meaningful interactions and collaborations enjoyed through the course of my career and pertaining to this book. These include (not in any particular order) CNS Centre for Neuro Skills; Dr Anis Rahman, Applied Research and Photonics; Professor R.G. Hunsperger, University of Delaware; Colonel (ret.) Dallas Hack, MD, US Army Medical Command; Dr. Zahra MirAfzali, Encapsula NanoSciences; Dr. Ramona Hicks, National Institutes of Health; Franklin and Marshall College Department of Physics; Professor Michael Mauk, Johns Hopkins University; U.S. Department of Veterans Affairs Office of Research and Development; Matt Willenkin, MD; Dr Dickron Mergerian, Westinghouse Advanced Technology Laboratory; Professor Jack Cramer, University of Delaware; Professor Larry Ezard, Penn State University; Dr Gary Vezzoli, Picatinny Arsenal; Professor Hermann Haus, MIT; Rob Kinsler, HP White Laboratory. I also thank the excellent production and editorial staff at CRC Press, Boca Raton, London, and New York.

DISCLAIMERS

The author does not endorse the articles, products, or guidelines referenced in this book. Information is provided to educate and raise awareness of the public health issues associated with mild traumatic brain injury and to provide a scientific overview of the topics. Patients should consult with medical experts for informed diagnosis and treatment and to obtain a comprehensive multidisciplinary evaluation, including physical examination with history of events and symptoms, neuroimaging, screening, and neuropathological testing.

Mild Traumatic Brain Injury: A Science and Engineering Perspective may include discussions on drugs or devices, or use of drugs or devices, that have not been approved by the Food and Drug Administration (FDA) or have been approved by the FDA for specific uses only. The FDA has stated that it is the responsibility of the physician to determine the FDA clearance status of each drug or device he or she wishes to use in clinical practice. The contents do not represent the views of the U.S. Department of Veterans Affairs or the U.S. Government.

Author

In the course of his career, Mark Mentzer developed a wide range of sensors and instrumentation. He operated an aerospace defense company; managed an optical-assembly manufacturing company; and led global-product development in Europe, Asia, and the Americas. He taught graduate engineering and physics at Penn State University and at the University of Maryland Global Campus and presented frequently at trade shows and conferences. His books are widely distributed in print and electronic media.

Mentzer contributed to the fields of neuroscience, biosensors, integrated optics, computer memory, aircraft fuel systems and safety, optical MEMS, terahertz frequency spectroscopy, X-ray cineradiography, fiber-sensor velocimetry, and high-brightness imaging. By applying his expertise in optical engineering, solid-state physics, biochemistry, and molecular biology, he authored more than a hundred papers and 17 patents. He may be reached through his website at www.markamentzer.com

OTHER BOOKS BY MARK A. MENTZER

Principles of Optical Circuit Engineering

Applied Optics Fundamentals and Device Applications Nano, MOEMS, and Biotechnology

Voyage of the Swamp Fox: A Trawler Cruising Adventure on the Intracoastal Waterway

Introduction

Mild traumatic brain injury (mTBI) presents a crisis in contact sports, the military, and public health. This book reviews our current understanding of mTBI, methods of diagnosis, treatment, policy concerns, and emerging technologies, all from the viewpoint of a scientist or an engineer. The intent is to address the scope of mTBI along with an understanding of triggering events. The broad coverage and accessible discussions will appeal to engineers and professionals in diverse fields related to mTBI, health care professionals, researchers working in athletic training, students of neurology, critical care medicine, neurology, neurosurgery, nurse practitioners, occupational therapy, pediatrics, physician assistants, psychiatry, psychology, sports medicine, trauma surgery, and military medicine, as well as policy makers and interested lay persons. Interdisciplinary engineers and scientists in the fields of neural engineering, biological and biomedical engineering, biomolecular engineering, systems biology, electrical engineering, signal and image processing will benefit.

Translational research forms the basis for the effective development of strategies and solutions in the mTBI discipline. Because of the widespread attention this topic achieves, there is a concerted attempt to address the Babylonian confusion of Gordian knots associated with the subject matter. However, none addresses the topic from the perspective of military-to-contact sports, insult-to-injury, bench-to-bedside, and diagnostics to therapies in the manner presented herein. This book serves as an entry point for those working in all aspects of the field of mTBI.

Recent understanding of the injury condition known as mild traumatic brain injury (mTBI), the medical condition chronic traumatic encephalopathy (CTE), certain aspects of post-traumatic stress disorders (PTSD) and their associated symptoms, as well as increased risk for other neuropsychiatric or neurodegenerative disorders such as Alzheimer's disease (Jellinger et al., 2001; Lye and Shores, 2000), further illuminate the need for improved understanding of the effects of brain injury. Improved prophylaxis includes armor designed to better shield from insult scenarios—as well as improved postexposure treatment to alleviate or minimize short- and longer-term effects of the insults. Academic research in TBI ranges from attempts to discover the pathophysiology, biomarkers, therapeutic targets, improved imaging modalities, and therapeutics and treatment strategies. A gamut of intracranial pathologies results in symptoms including loss of memory, disorientation, and long-term cognitive disorders (Hoffman, 2015).

An estimated 1.6–3.8 million TBI cases occur annually in the United States (McKee and Daneshvar, 2015). The majority of cases are closed-head mTBI injuries. Reported incidence is on the rise, for instance 62% increase in mTBI related to recreation over 10 years (Coronado et al., 2015). The CDC estimates 2.2 million emergency department visits from TBI associated with 50,000 deaths annually; and the DoD reports TBI diagnoses for 4.2% of those in the services for the first 11 years of this century (Centers for Disease Control Prevention, 2014). The Defense and Veterans Brain Injury Center (DVBIC) estimates 1.7 million annual TBIs in the United States, of which 84% are mTBIs (DVBIC, 2015). The World Health Organization (WHO) predicted that TBIs will represent the third-leading cause of mortality and disability globally by the year 2020 (World Health Organization, 2009). Direct medical costs plus economic impacts such as lost productivity related to brain injuries in the United States are estimated at more than $60 billion annually (Finkelstein et al., 2006).

It is now understood the brain may not fully recover with time. Second impact syndrome relates the heterogeneous response to repetitive head impacts during the vulnerable period before resolution of the initial impact has occurred. Several groups established guidelines for classification of a head injury as a mTBI (DeWitt et al., 2013). These include the Department of Veterans Affairs, the Department of Defense Clinical Practice Guidelines, and the World Health Organization. According to the guidelines a head injury is classified as a mTBI if there is loss of consciousness for less than 30 minutes; alteration of consciousness less than 24 hours; post-traumatic amnesia of less than a day; and an initial Glasgow Coma Scale of 13–15. The Center for Disease Control (CDC) and National Institutes of Health (NIH) advocate research to fully understand mTBI, risk factors, and strategies to reduce incidence and improve outcomes.

The American National Football League set out major initiatives subsequent to drastically increased recognition and focus on the plethora of head injuries resulting from harmful collisions on the field of play. Video capture and replay from multiple angles around the field allow slowing, zooming, and tagging footage relevant to potential injuries. Replay footage is used to effect immediate identification, medical treatment, and analysis of injury conditions. This is provided to team physicians and trainers with direct communication to referees to stop game play in possible injury scenarios. Further analysis of injuries and relevant conditions helps officials and the league in making future rule changes and safety improvements.

We begin by detailing the epidemiology and neurophysiology of mTBI, presenting disease models and descriptions of nucleating events. This is followed by characterization of sensors, imagers, and related diagnostic measures for both injury analysis and medical diagnosis. Relationships are drawn with emerging bioinformatics analysis and mTBI markers. We continue with discussions of sports medicine and military issues, finally covering therapeutic strategies, drug trials and candidates, and future developments.

The field of clinical neurology needs a detection system to identify and quantify the extent of trauma an individual receives and to allow for rapid-treatment decision making in the field or in a clinical setting. Secondary injury to central nervous system tissue associated with the physiologic response to an initial insult resulting from athletic injury, direct blunt force or the percussive forces found in close proximity to a military environment blast source could be lessened if the initial insult could be rapidly diagnosed or characterized (Mentzer, 2013).

While the diagnosis of severe forms of such insults and the resulting damage is straightforward through clinical response testing, and computed tomography (CT) and magnetic resonance imaging (MRI) testing, these diagnostics have their limitations for assessment of mTBI. Medical imaging is both costly and time-consuming, while clinical response testing of incapacitated individuals is of limited value and often precludes a nuanced diagnosis. In many instances, the instrumentation necessary for these diagnostic procedures is not available in the field.

Additionally, owing to the limitations of existing diagnostics, situations exist under which subjects experience a stress to their neurological condition such that the subject often is unaware that damage has occurred or does not seek treatment as the subtle symptoms often quickly resolve. The lack of treatment of mild to moderate challenges to neurologic conditions can have a cumulative effect or subsequently result in a severe brain damage event that has a poor clinical prognosis. Hence work continues in development of effective test metrics, associated diagnostic criteria, and supporting measurement instrumentation (Mentzer, 2015).

Analysis of the mechanisms and development of biomarkers related to mTBI is complicated by a deficiency in experimental studies and by lack of sensitivity and specificity of biomarker-based injury prediction. By the time a biomarker analysis is performed the subject may already be in a severe and irreversible state. Thus, there is a need for a detection system that can identify the presence or absence of an event severe enough to warrant monitoring or treatment and optionally quantify the extent of trauma an individual has received that will allow for rapid treatment and decision making in the field or in a clinical setting.

REFERENCES

Centers for Disease Control Prevention; National Center for Injury Prevention and Control; Division of Unintentional Injury Prevention. 2014. *Report to congress on traumatic brain injury in the United States: Epidemiology and rehabilitation.* Atlanta, GA: Centers for Disease Control Prevention.

Coronado, V.G., S.T. Haileyesus, T.A. Cheng, et al. 2015. Trends in sports- and recreation-related traumatic brain injuries treated in US emergency departments: the national electronic injury surveillance system-all injury program (NEISS-AIP) 2001–2012. *J Head Trauma Rehabil.* 30:185–197.

Defense and Veterans Brain Injury Center. 2015. *About traumatic brain injury.* Silver Spring, MD: DVBIC.

DeWitt, D.S., R. Perez-Polo, C.E. Hulsebosch, P.K. Dash, and C.S. Robertson. 2013. Challenges in the development of rodent models of mild traumatic brain injury. *J Neurotrauma* 30:688–701.

Finkelstein, E.A., P.S. Corso, T.R. Miller, et al. 2006. *The incidence and economic burden of injuries in the United States.* New York: Oxford University Press.

Hoffman, S.W., in forward to F.H. Kobeissy, ed. 2015. *Brain neurotrauma-molecular, neuropsychological, and rehabilitation aspects.* Boca Raton: CRC Press.

Jellinger, K.A., W. Paulus, C. Wrocklage, and I. Litvan. 2001. Traumatic brain injury as a risk factor for Alzheimer disease. Comparison of two retrospective autopsy cohorts with evaluation of ApoE genotype. *BMC Neurol.* 1:3.

Lye, T.C. and E.A. Shores. 2000. Traumatic brain injury as a risk factor for Alzheimer's disease: a review. *Neuropsychol Rev.* 10:115–129.

McKee, A.C. and D.H. Daneshvar. 2015. The neuropathy of traumatic brain injury. *Handb Clin Neurol.* 127:45–66.

Mentzer, M.A. 2013. *Analysis and design of a photonic biosensor for mild traumatic brain injury.* Aberdeen: Army Research Laboratory.

Mentzer, M.A. 2015. *Blast, ballistic and blunt trauma sensor.* Patent US9080984, Alexandria: United States Patent Office.

World Health Organization. 2009. *Global health risks: mortality and burden of disease attributable to selected major risks.* Washington, DC: World Health Organization.

1 Clinical Sequelae and Functional Outcomes

DEFINITIONS OF mTBI

Following are some common definitions for TBI. These provide an overview of the concerns and varied stakeholder perspectives.

CENTERS FOR DISEASE CONTROL

A traumatic brain injury is caused by a bump, blow, or jolt to the head or a penetrating head injury that disrupts the function of the brain. (CDC, 2016)

NATIONAL INSTITUTE OF NEUROLOGICAL DISORDERS AND STROKE

TBI, a form of acquired brain injury, occurs when a sudden trauma causes damage to the brain. TBI can result when the head suddenly and violently hits an object, or when an object pierces the skull and enters brain tissue. (NINDS, 2012)

DEPARTMENT OF DEFENSE

A traumatically induced structural injury or physiological disruption of brain function as a result of external force that is indicated by new onset or worsening of at least one of the following clinical signs, immediately following the event:

- any alteration in mental status (e.g., confusion, disorientation, slowed thinking)
- any loss of memory for events immediately before or after the injury
- any period of loss or a decreased level of consciousness, observed or self-reported.

External forces may include any of the following events: the head being struck by an object, the head striking an object, the brain undergoing an acceleration/deceleration movement without direct external trauma to the head, or forces generated from events such as blast or explosion, including penetrating injuries. (Dept. of Defense, 2014)

AMERICAN PSYCHIATRIC ASSOCIATION

TBI is the injury sustained and can be the nucleating factor for a host of neurological disorders. Neurocognitive disorder (NCD) due to TBI is termed a DSM-5 diagnosis regarding cognitive symptoms of impairments following a TBI. About 80% of cases are mild, 10% moderate, and 10% severe, but "at a minimum, TBI produces a diminished or altered state of consciousness. TBI results in a diverse, idiosyncratic constellation of cognitive, neurological, physical, sensory, and psychosocial symptoms" (American Psychiatric Association, 2013).

NATIONAL ACADEMY OF SCIENCES

The National Academy of Sciences (National Academies of Science, Engineering, and Medicine, 2019) defines traumatic brain injury as

> an insult to the brain from an external force that leads to temporary or permanent impairment of cognitive, physical, or psychosocial function. TBI is a form of acquired brain injury, and it may be open (penetrating) or closed (non-penetrating) and can be categorized as mild, moderate, or severe, depending on the clinical presentation.

BRAIN INJURY ASSOCIATION OF AMERICA

> TBI: "alteration in brain function, or other evidence of brain pathology, caused by an external force."
>
> Acquired brain injury (ABI): "injury to the brain which is not hereditary, congenital, degenerative, or induced by birth trauma … an injury to the brain that has occurred after birth."

AMERICAN CONGRESS OF REHABILITATION MEDICINE

> A patient with mild traumatic brain injury is a person who has had a traumatically induced physiological disruption of brain function as manifested by at least one of the following (Ashley and Hovda, 2018; Kay et al., 1993)

- Any period of loss of consciousness
- Any loss of memory for events immediately before or after the accident
- Any alteration in mental state at the time of the accident (e.g., feeling dazed, disoriented, or confused)
- Focal neurological deficit(s) that may or may not be transient, but where the severity of the injury does not exceed the following:
  - Loss of consciousness of approximately 30 minutes or less
  - After 30 minutes, an initial Glasgow Coma Scale (GCS) of 13–15
  - Posttraumatic amnesia (PTA) not greater than 24 hours

CARNEY ET AL. DEFINITION

> Carney et al. (2014) provided a definition:

- A change in brain function after a force to the head that may be accompanied by temporary loss of consciousness
- Indicators of concussion, identified in an alert individual after a force to the head that include the following:
  - Observed and documented disorientation or confusion immediately after the event

- Slower reaction time within 2 days postinjury
- Impaired verbal learning and memory within 2 days postinjury
- Impaired balance within 1 day postinjury

AMERICAN ASSOCIATION OF NEUROLOGICAL SURGEONS

The AANS (2020) defined concussion as "A clinical syndrome characterized by immediate and transient alteration in brain function, including alteration of mental status and level of consciousness, resulting from mechanical force or trauma."

CSTE AND BU SCHOOL OF MEDICINE

The Center for the Study of Traumatic Encephalopathy (CSTE) and Boston University School of Medicine defined chronic traumatic encephalopathy (CTE) as:

> a progressive neurodegenerative disease caused by repetitive trauma to the brain … characterized by the build-up of a toxic protein called Tau in the form of neurofibrillary tangles (NFT's) and neuropil threads (NT's) throughout the brain. The abnormal protein initially impairs the normal functioning of the brain and eventually kills brain cells. Early on, CTE sufferers may display clinical symptoms such as memory impairment, emotional instability, erratic behavior, depression and problems with impulse control. However, CTE eventually progresses to full-blown dementia.

CONCUSSION IN SPORT CONFERENCE

The Fourth Annual Conference on Concussion in Sport (McCrory et al., 2013) provided a consensus statement:

> Concussion is a brain injury and is defined as a complex pathophysiological process affecting the brain, induced by biomechanical forces. Several common features that incorporate clinical, pathologic, and biomechanical injury constructs that may be utilized in defining the nature of a concussive head injury include

> 1. Concussion may be caused either by a direct blow to the head, face, neck, or elsewhere on the body with an "impulsive" force transmitted to the head.
> 2. Concussion typically results in the rapid onset of short-lived impairment of neurological function that resolves spontaneously. However, in some cases, symptoms and signs may evolve over a number of minutes to hours.
> 3. Concussion may result in neuropathological changes, but the acute clinical symptoms largely reflect a function disturbance rather than a structural injury and, as such, no abnormality is seen on standard neuroimaging studies.
> 4. Concussion results in a graded set of clinical symptoms that may or may not involve loss of consciousness. Resolution of the clinical and cognitive symptoms typically follows a sequential course. However, it is important to note that, in some cases, symptoms may be prolonged.

NCAA

The National Collegiate Athletic Association (NCAA) defines concussion or mild traumatic brain injury as "a complex pathophysiological process affecting the brain, induced by traumatic biomechanical forces."

In each of the definitions the forces applied to the head may result from forces applied elsewhere and transmitted through the body to the head. Each of the definitions states or implies that symptoms may not necessarily be transient. Factors specific to each patient may contribute to disposition toward a response to mechanical insult, thereby complicating prognostication. Most individuals experience resolution of symptoms within 90 days of injury (Alexander, 1995). There is also lack of understanding of the relationship between activity and rest levels and how these factors influence symptom persistence (Ashley and Hovda, 2018).

WORLD HEALTH ORGANIZATION (WHO, 1995)

A case of traumatic brain injury (craniocerebral trauma) is either:

- An occurrence of injury to the head (arising from blunt or penetrating trauma or from acceleration–deceleration forces) with at least one of the following:
  - Observed or self-reported alteration of consciousness or amnesia due to head trauma
  - Neurologic or neuropsychological changes or diagnoses of skull fracture or intracranial lesions that can be attributed to the head trauma
- Or an occurrence of death resulting from trauma with head injury or traumatic brain injury listed on the death certificate, autopsy report, or medical examiner's report in the sequence of conditions that resulted in death

Examples of neurologic changes include abnormalities of motor function, sensory function, or reflexes; abnormalities of speech (aphasia or dysphasia); or seizures acutely following head trauma. Examples of neuropsychological abnormalities include amnesia as described previously; agitation or confusion; and other changes in cognition, behavior, or personality.

SEVERITY CLASSIFICATION OF TBI AND SCREENING TOOLS

The severity of TBI is classified using the Glasgow Coma Scale (GCS), Loss of Consciousness (LOC), and post-traumatic amnesia (PTA), along with a variety of other screening tools such as ANAM (Automated Neuropsychological Assessment Metrics), the Repeatable Battery for Assessment of Neuropsychological Status, the Concussion Management Algorithm (CMA), the King-Devick concussion test (North et al., 2012; Marshall et al., 2012), the Sport Concussion Assessment Tool (SCAT3), and the Acute Concussion Evaluation (Ontario Neurotrauma Foundation, 2013; Gioia and Collins, 2006). And no single classification embraces all the features of mTBI (clinical, pathological, cellular/molecular). The severity does not directly equate to neurocognitive disorder (NCD) or the potential for rehabilitation.

Many factors such as injury specifics, age, prior history, and substance abuse relate to the effects of a TBI (Relias Academy, 2020). Edition 5 of the *Diagnostic and Statistical Manual for Mental Disorders* (DSM-5) describes the neurocognitive sequelae following TBI. NCD encompasses the group of acquired disorders wherein the primary clinical deficit is disrupted cognitive functioning (American Psychiatric Association, 2013). DSM-5 is the standard classification of mental disorders used by mental health professionals in the United States.

Symptoms often associated or comorbid with mTBI are often classified as acute, subacute, or late as follows (Relias Academy, 2020)

- Acute loss of consciousness, amnesia, headache, confusion, nausea, vomiting, vertigo, tinnitus, changes in vision
- Subacute problems with cognition and memory, disturbed sleep, irritability, fatigue
- Late epilepsy, NCD due to Alzheimer's, or Parkinson's disease

While the relationships of disease states and associated disorders to mTBI are not well understood, post-concussive symptoms (physical, cognitive, emotional/behavioral) are easily overlooked or misdiagnosed as mental health or other physical problems. For instance, PTSD and TBI share symptoms, making a differential diagnosis difficult. With both conditions present, the symptoms can be mutually exacerbating. In recent wars, comorbidity was 48% (American Psychiatric Association, 2013). Another possible comorbidity is depression. Figure 1.1 (Courtesy of CNS Centre for Neuro Skills) illustrates skills and functions associated with the lobes of the brain along with changes observed after brain injury.

GLASGOW COMA SCALE

The Glasgow Coma Scale (GCS) (Teasdale et al., 2014) describes the level of consciousness following mTBI and determines the level of severity of the injury GCS measures the following:

Eye Opening (E)
- 4 = spontaneous
- 3 = to sound
- 2 = to pressure
- 1 = none
- NT = not testable

Verbal Response (V)
- 5 = oriented
- 4 = confused
- 3 = words, but not coherent
- 2 = sounds, but no words
- 1 = none
- NT = not testable

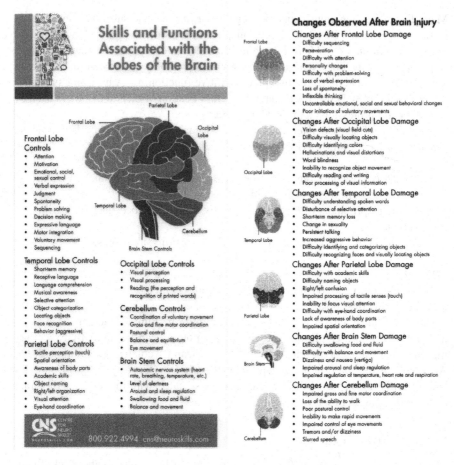

FIGURE 1.1 Skills and functions associated with the lobes of the brain and changes observed after brain injury (Courtesy CNS Centre for Neuro Skills cns@neuroskills.com).

Motor Response (M)

- 6 = obeys command
- 5 = localizing
- 4 = normal flexion
- 3 = abnormal flexion
- 2 = extension
- 1 = none
- NT = not testable

The GCS score is the sum of the assessment scores. There is a pediatric version (PGCS).

Screening for mTBI in sports often involves the use of the Standardized Assessment of Concussion (SAC) or the Mini Mental Status Exam (MMSE) (Brennan et al., 2014). Military screening uses the Military Acute Concussion Evaluation (MACE) tool. This screening is most suitable for screening within 12 hours after an injury, and includes documentation and brief cognitive and neurological functions elements (Assistant Secretary of Defense, 2015).

RANCHO LOS AMIGOS SCALE

The Rancho Los Amigos Scale, often used with severe TBI, determines the severity of a TBI, measuring patient cognition, awareness, behavior, and interaction to arrive at one of eight levels of responsiveness to external stimuli as follows (Arrowhead Publishers, 2014):

Level 1—no response to visual, verbal, tactile, auditory, noxious stimuli
Level 2—generalized response
Level 3—localized response
Level 4—confused–agitated
Level 5—confused–inappropriate
Level 6—confused–appropriate
Level 7—automatic–inappropriate
Level 8—purposeful and appropriate
Level 9—purposeful and appropriate (standby assistance on request)
Level 10—purposeful and appropriate (modified independent)

WESTMEAD POST-TRAUMATIC AMNESIA (PTA) SCALE

Duration of post-traumatic amnesia is a useful indicator of TBI. The PTA scale uses questions and flash cards to assess patient alertness. A perfect score for three consecutive days indicates lack of PTA (Arrowhead Publishers, 2014).

Prognosis and rehabilitation potential may be inferred from the patient's TBI classification. Complications may develop at all levels of classification; therefore, accurate diagnosis is very important. A disability rating scale (DRS) may overcome some of the subjective limitations of other measures, as the objective is more sensitive and uses quantitative measurement variables. In the DRS, points are assigned for arousability—eye opening, communication ability, motor response; cognitive ability for self-care activities—feeding, toileting, grooming; dependence on others—level of functioning; and psychological adaptability—employability (Arrowhead Publishers, 2014).

INTERNATIONAL MISSION FOR PROGNOSIS AND CLINICAL TRIAL (IMPACT) PROGNOSIS MODEL

The IMPACT prognostic model and calculator is available for outcome predication. The model is based on integration of observations, TBI variables, therapeutic studies, common data elements, and trial outcomes. According to their website (http://tbi-impact.org/), the IMPACT Project is focused on advancing knowledge of prognosis, trial-design, and treatment in TBI. IMPACT has

- developed and validated prognostic models for classification and characterization of TBI series
- participated in the development of standardization of data collection in TBI studies (common data elements)
- provided evidence-based recommendations for improving sensitivity and efficiency of trials in TBI

Collaboration with the Medical Research Council's (MRC's) corticosteroid randomization after significant head injury (CRASH) prognosis model makes the combined study the largest performed to date (Arrowhead Publishers, 2014). This work represents an important decision-support system for medical doctors and their patients. TBI prognosis is quite challenging, and numerous complications are associated with TBI.

Complications include neurological, physical, cognitive, and emotional/behavioral issues, often collectively termed post-concussion syndrome (PCS). CTE from mTBI is linked with other delayed neurodegenerative diseases, including Alzheimer's, Parkinson's, and amyotrophic lateral sclerosis (Lou Gehrig's disease) (Arrowhead Publishers, 2014). While accumulations of hyperphosphorylated tau are directly associated with the term CTE, a number of conditions are associated with high levels of cerebral tau aggregation; and there no firmly established clinical or pathological criteria for the diagnosis of CTE (Ashley and Hovda, 2018).

LIMBIC-CENC Consortium

Yet another set of tools for concussion assessment is available from the Long-term Impact of Military-Relevant Brain Injury Consortium-Chronic Effects of Neurotrauma Consortium (LIMBIC-CENC, 2020). According to their website (Walker et al., 2016),

> A multilayered structured interview was developed for the Chronic Effects of Neurotrauma Consortium (CENC) to standardize the process of collecting and classifying self-report information on every potential concussive event (PCE) throughout lifetime. It consists of a 'mapping' interview that systematically identifies all events during lifetime that may have resulted in a TBI. Each event is further queried with either in-depth TBI diagnostic interview or brief description of the event followed by several TBI screening questions.

Boston Assessment of Traumatic Brain Injury-Lifetime (BAT-L)

According to the VA website (VA research 2019) researchers developed an assessment tool to diagnose and evaluate traumatic brain injury in Veterans from the recent wars in Iraq and Afghanistan. The Boston Assessment of Traumatic Brain

Injury-Lifetime (BAT-L) has become a gold standard for assessing mild TBI specific to Veterans. It is in use at many VA medical centers and War Related Illness and Injury Study Center sites across the country.

The VA Translational Research Center for TBI and Stress Disorders (TRACTS) developed the BAT-L. It is a semi-structured clinical interview to characterize head injuries and diagnose TBIs throughout a Veteran's lifespan. Unlike existing TBI assessments, the BAT-L probes the unique experiences of combat-exposed Veterans, such as possible repeated blast exposures. The BAT-L evaluates possible TBIs over the lifetime of a Veteran, rather than just immediately after a traumatic event. A civilian variation on the interview has also been developed.

mTBI PATHOPHYSIOLOGY

While the exact pathology of TBI is not well understood, dynamic pathophysiology evolves with time and involves into two major stages. Initial trauma produces systemic local cerebral environmental changes that cause secondary injury. Secondary injury involves numerous molecular pathways (Rosenfeld et al., 2012). Cascading intracellular biochemical imbalances cause uncontrolled release of neurotransmitters, calcium overload, inflammation, oxidative damage, and hyperactivity of enzymes, leading to cell death (Arrowhead Publishers, 2014).

Transient forces on the order of milliseconds damage cellular membranes, axons, and vasculature, producing histopathological indicators of mTBI in multiple regions of the brain. Diffuse effects of trauma involve changes to material properties of tissues, such as gray–white matter interfaces, axon hillocks, and blood-brain barrier (Farkas and Povlishock, 2007). Mechanical shearing of axons, vasculature, and membranes promotes cellular signaling such as leukocyte infiltration and exposure to circulating and endogenous cytokines and chemokines (Gentleman et al., 2004). Ionic redistribution raises extracellular potassium, depolarizes membranes, and releases neurotransmitters, while elevated glutamate and other excitatory amino acids trigger progression to cellular damage and death, axotomy, and synaptic deafferentation (Blinzinger and Kreutzberg, 1968) and dysfunction of the neurovascular unit (Farkas and Povlishock, 2007). The posttraumatic response of the brain also includes a responsive neuroplastic regenerative change that varies with brain region (Lifshitz, 2015).

As described by (Joseph, 2011),

Characteristically, following head trauma there is an immediate and global reduction in cardiac output and blood flow for a few seconds, which in turn results in oxygen deprivation. This is followed by a transient hypertension and then a prolonged hypotension. The overall consequences are reductions in arterial blood perfusion and thus hypoxia throughout the brain as well as decreased blood flow within the damaged tissue. Metabolism is disrupted and energy failure occurs. In addition, the brain's blood and oxygen supply can be reduced or diverted following chest injury, scalp laceration, or fractures involving the limbs.

Unfortunately, a vicious deteriorative feedback cycle can be produced by these conditions which further reduces the brain's supply of blood and oxygen. For example, hypoxia induces vasodilation and constriction of the blood vessels, a condition which results in intracranial hypertension. This creates an overall increase in intracranial pressure. When intracranial pressure increases the brain and blood vessels become compressed (due to displacement pressure) which further reduces blood and oxygen flow within the brain and to the damaged tissue. This causes the development of ischemia and focal cerebral edema, all of which in turn add to displacement pressures (herniation).

Hence, a variety of conditions associated with head injury can conspire to produce widespread hypoxia and disturbances involving cerebral metabolism. Neuronal death, independent of impact or rotational acceleration-deceleration forces is the long term consequence, even in mild cases.

Astrocytes take up glutamate, convert it to glutamine, and deliver it to neurons as an energy source. After TBI, cell death produces excess glutamates, which bind to N-methyl-D-aspartate (NDMA) neuronal receptors, inducing an influx of calcium and sodium and efflux of potassium in and out of the neuron. The resulting ionic imbalance depolarizes the cell membrane and produces an overload of intracellular calcium. This leads to mitochondrial impairment, decreased adenosine triphosphate (ATP) formation, and cell death (Rosenfeld et al., 2012).

Mitochondrial impairment also causes release of reactive oxygen and nitric oxide, leading to oxidative stress damage to cellular membrane lipids, proteins, and DNA, and impairment of axonal transport and function due to the disruption of axonal cytoskeletal filaments (Rosenfeld et al., 2012). In addition, glial cell activation secretes inflammatory cytokines, chemokines, and neurotrophins, and phagocytosis by microglia leads to swelling and blood-brain barrier impairment, along with increased intracranial pressure and ischemia (Rosenfeld et al., 2012).

Subsequent to TBI, cerebral edema, or swelling of the brain, leads to neuronal damage and disrupts blood and oxygen flow (anoxia). If severe, this can compress the brain stem and result in the death of the patient. With sufficient damage to blood vessels, a pool of blood, termed a hematoma, may develop. This creates increased intracranial pressure and further brain damage, often necessitating drainage by the neurosurgeon (Brain Injury Association of America, 2012).

Each brain injury is unique and, depending on the brain locations injured, produces a unique combination of disabilities. For injury to the left side of the brain, symptoms may include impairment of receptive and/or expressive language, depression and anxiety, verbal memory deficits, decreased control of the right side of the body, impaired logic, and sequencing difficulties. Injuries to the right side of the brain may include impaired visual-spatial perception, decreased control of the left side of the body, altered creativity and music perception, loss of the big picture (Gestalt), and visual memory deficit. With diffuse injury, impairments may include reduced thinking speed, increased confusion, reduced attention and concentration, increased fatigue, and impaired cognitive function (Brain Injury Association of America, 2012).

Depending on the severity and repetition of the mTBI presentation, a range of cascaded intracranial pathologies presents. Pathologies include diffuse axonal injury, mechanical-tissue damage, ischemia, synaptic loss, and neuronal dysfunction or demise. These relate to tissue damage and cellular biochemical pathway disruptions. Attributes include impaired axonal transport, neuronal-circuit disruption, contusions, mild edema, variable chronic cognitive or neuropsychiatric impairments, and post-traumatic stress disorder (PTSD).

There is also a cumulative effect, where repeated insults produce axonal and cyto-skeletal alterations. An important effect is the "apolipo" protein e4 allele expression resulting in reduced abnormal tau clearance and consequent tauopathies, including beta amyloid plaques and neurofibrillary tangles, which can spread among cells through anatomical connections (Soto, 2012; De Calignon et al., 2012). Repeated mTBI eventually culminates in what was called "retrograde amnesia" in boxers, but what is now understood to be a group of neurological disorders marked by hypoki-nesia, tremor, and muscular rigidity—or Parkinsonism—and the key impact is now collectively referred to as reduced neural plasticity.

Looking at the physiological processes for mTBI we see positive feedback loops that enhance and amplify biochemical pathway disruptions. mTBI results in what is termed a "neurotransmitter storm" of disruptions that are interrelated pathologies over time. TBI results in an uncoupling of blood flow and metabolism, evidenced in either insufficient blood flow to the brain (aka ischemia) or increased blood flow and diffuse swelling (hyperemia). It is the former (cerebral hypoperfusion) that is most frequently observed in cases of TBI. Since the mechanisms for TBI mediation are not clearly established, potentially interrelated studies are of great importance for exploration of cerebral hemodynamics following TBI.

Figure 1.2 (North et al., 2012) illustrates the pathophysiology of brain injury, including shear, lacerations, and tearing of cellular membranes and tissue structures, along with extreme cases of diffuse axonal injury to axon structures and myelin sheaths. Figure 1.3 (North et al., 2012) illustrates physiological changes correlated to the pathophysiologies of Figure 1.2.

At the molecular level, short-term memory is attained by modifying the short-term plasticity at the synapse through synaptic facilitation, potentiation, and aug-mentation. These actions change the dynamics of transmission at synapses as a result of recent synaptic activity. A step-by-step example is when a high frequency series of presynaptic action potentials create

1. post synaptic facilitation due to Ca^{2+} rushing in but not leaving quickly, prolonging the elevation
2. followed by synaptic augmentation due to, again, the rise of Ca^{2+}
3. synaptic depression due to depleted neurotransmitter vesicles in the presyn-aptic neuron
4. potentiation—again thought to result in elevated levels of Ca^{2+}

These effects are temporary—hence thought to be the cause of short-term memory. Long-term memory is when the physiology of the neurons actually changes, yield-ing differential mechanical structures, hence enabling long-term memories. This is similar to changing the way a circuit board works when adding more transistors, resistors, etc., fundamentally changing the way signals are processed. Long-term memory effect is achieved by continuous long-term potentiation (which are ini-tially created by NMDA receptor losing the Mg^{2+} block, allowing Ca^{2+} rush with

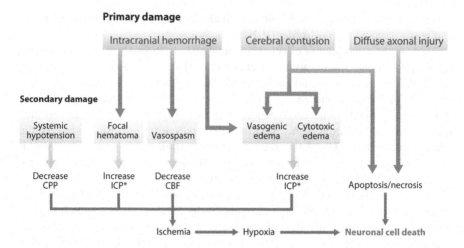

FIGURE 1.2 Pathophysiology of brain injury. *Abbreviations*: CBF, cerebral blood flow; CPP, cerebral perfusion pressure; ICP, intracranial pressure. The asterisks indicate the presence of ICP biomarkers C-tau, GFAP, and S100β. Adopted from North et al. (2012) *Annu. Rev. Anal. Chem.* 5:35–56.

the condition that action potential originates at the presynaptic neuron and not the postsynaptic neuron) resulting in transcriptional activity coding additional receptors, synapse junctions, and proteins that change neuron operating dynamics. Figure 1.4 illustrates the basic structure and function of the neuron including the myelin sheath and neurotransmission.

Eric Kandel's Nobel winning work (Kandel, 2006) focused on the model organism aplysia (large sea slug). He found the rapid gill-withdrawal reflex, controlled by the approximately 20,000 nerve cells in the abdominal ganglion, could be modified by habituation and sensitization—each of which give rise to short-term memory for a few minutes. The protective reflex when the slug is touched lessens with repeated touches as the slug realizes the response is trivial (habituation). Sensitization was accomplished via electrical shock—where the slug exaggerated the reflex response to touch, even in the absence of shock.

Kandel next created long-term memory by repeated training, interspersed with rest periods—and equated this with long-term memory training in mammals. His objective was to identify evolutionarily conserved learning modalities, so that the relatively simple sea slug processes could provide the basis for identifying the cellular basis for learning in the mammalian hippocampus. He hoped to discern whether neural circuits were unique, or if the connections were random and equally valuable to the learning processes. And he hoped to address the conversion of short-term memory to long term. Long-term sensitization in the slug may be limited to simple sensitization, where long-term potentiation (LTP) in mammals involves evolved processes of synaptic plasticity, Hebb's postulate (Bear and Malenka, 1994), and memory storage. This process is further distinguished in mammals vs. sea slugs by

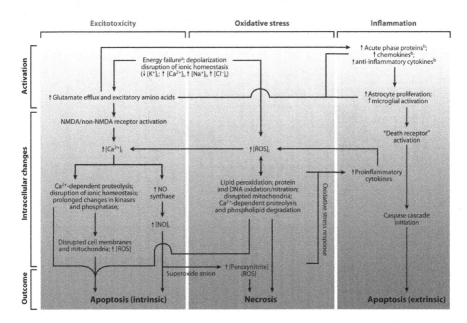

FIGURE 1.3 Biochemical cascades following brain injury. The up arrow indicates an increase or upregulation; the down arrow indicates a decrease or downregulation. The brackets with subscript i enclose an intracellular element. Superscript a refers to energy failure ($\uparrow$ glycolysis, $\downarrow$ ATP, $\downarrow$ pH); superscript b refers to acute-phase proteins [e.g., C-reactive protein, amyloid A, interleukin (IL)-1, IL-6, tumor necrosis factor α], anti-inflammatory cytokines (e.g., IL-10, transforming growth factor β), and chemokines [e.g., intercellular adhesion molecule 1, macrophage inflammatory protein (MIP)-1, MIP-2]. *Abbreviations*: NMDA, *N*-methyl-D-aspartic acid; NO, nitric oxide; ROS, reactive oxygen species. Adopted from North et al. (2012) *Annu. Rev. Anal. Chem.* 5:35–56.

the coincidence timing, orientations for spatial and temporal components of specific mammalian memory, and associativity.

Focusing on one aspect of these behaviors, LTP described the long-term improved synaptic strength, or efficacy, and the cellular and molecular basis that produces LTP at mammalian (hippocampus as opposed to the cerebellum effects described elsewhere) brain synapses (or sea slug ganglia). Perhaps the short- and long-term comparisons of the sea slug vs. mammalian behavior differ depending on the part of the brain producing the response—for example, the amygdala (more sea slug like process) vs. the mammalian hippocampus long-term memory retrieval as the basis for a "more considered" response.

It will be interesting to distinguish signaling pathways for LPT of Shaffer collaterals from the NMDA receptor responses to further consider the permutations of pre- and post-synaptic temporal relationships to LTP. The late LTP may be distinguished from early LPT, where the former involves CREB-activated transcription, triggered by protein kinase activators (see Appendix 2).

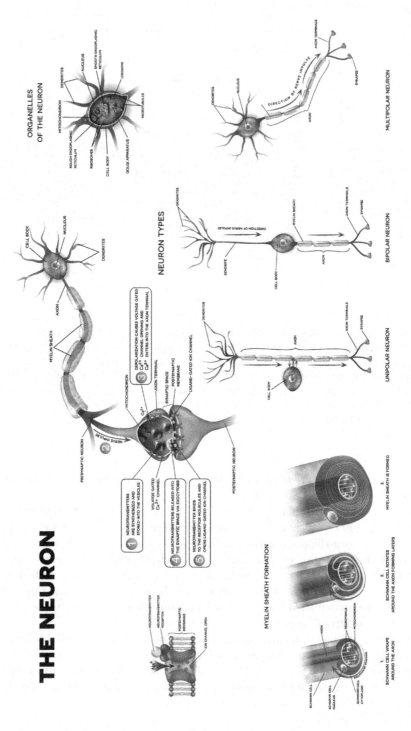

FIGURE 1.4 Structure and function of the neuron.

DIAGNOSIS AND DECISION-MAKING

The VA/DoD Clinical Practice Guideline for the Management of Concussion-Mild Traumatic Brain Injury (VA/DoD, 2016) assists decision-making to those defining a standard of care. The guidelines are based on a systematic review of clinical and epidemiological evidence. The stated expected outcomes for the guideline are to:

- Assess the patient's condition and determine the best treatment method
- Optimize the clinical management to improve symptoms and functioning adherence to treatment, recovery, well-being, and quality of life outcomes
- Minimize preventable complications and morbidity
- Emphasize the use of patient-centered care

The VA/DoD clinical guidelines provide a classification of TBI severity from mild to moderate to severe. Regarding loss of consciousness (LOC), 0–30 minutes = mild; >30 minutes and <24 hours = moderate; and >24 hours = severe. For alteration of consciousness/mental state (AOC), up to 24 hours = mild; and >24 hours may be moderate or severe. Posttraumatic amnesia (PTA) of 0–1 day = mild; >1 day and <7days = moderate; and >7 days = severe. Glasgow Coma Scale (GCS) scores are best obtained in the first 24 hours (use best score). 13–15 = mild; 9–12 = moderate; and <9 = severe. Note that in April 2015 a DOD memorandum recommended against the use of the GCS score for TBI diagnosis (Fortier et al., 2014, 2015).

In the classification notes it is stated,

Alteration of mental status must be immediately related to the trauma to the head. Typical symptoms would be looking and feeling dazed and uncertain of what is happening, confusion, and difficulty thinking clearly or responding appropriately to mental status questions, and being unable to describe events immediately before or after the trauma event.

Figures 1.5–1.7 delineate the VA/DoD algorithms intended to facilitate clinical diagnostic and therapeutic decision-making for mTBI management (adapted from VA/DOD 2009 and 2016. *Clinical practice guideline for the management of concussion–mild traumatic brain injury version 2.0.* Department of Veterans Affairs and Department of Defense.). Rounded rectangles present clinical states and conditions. Hexagons are decision points and rectangles are actions taken during the care process.

ETIOLOGY OF mTBI

Kinetic energy is transferred to a player's body from an unyielding surface during an impact due to biomechanical loading. In impulsive biomechanical loading, kinetic energy is transferred indirectly due to changes in motion via acceleration and deacceleration. The head is the most vulnerable body part to biomechanical loading and the brain is the most vulnerable organ.

Biomechanical loading causes shearing forces in the brain leading to concussions. The head and neck better accommodate linear front-to-back nodding movements than sideways movements in a rotational mode, but many football loadings to the

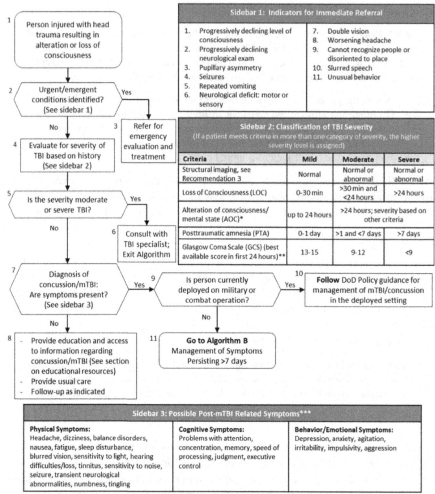

FIGURE 1.5   Initial presentation (>7 days post-injury). Adapted from VA/DOD (2009, 2016).

head involve angular acceleration–deceleration of 138 *g* lasting up to 15 ms (Omalu, 2008).

Blast from IEDs is the most common cause of injury in the battlefield. There are four classes of blast injuries (Brainline, 2020).

1. primary blast caused by shock waves traveling through the body
2. secondary blast from shrapnel and other flying debris

3. tertiary blast occurs when a service member is accelerated into a vehicle's interior or to the ground
4. quaternary blast is caused by crushing, burning, or inhalation of smoke and gases.

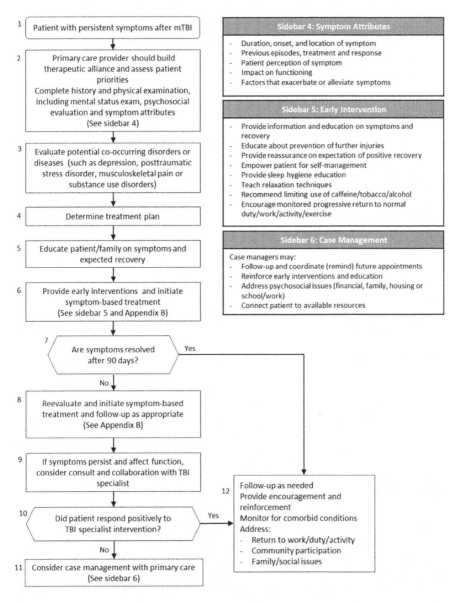

FIGURE 1.6 Management of symptoms persisting >7 days. Adapted from VA/DOD (2009, 2016).

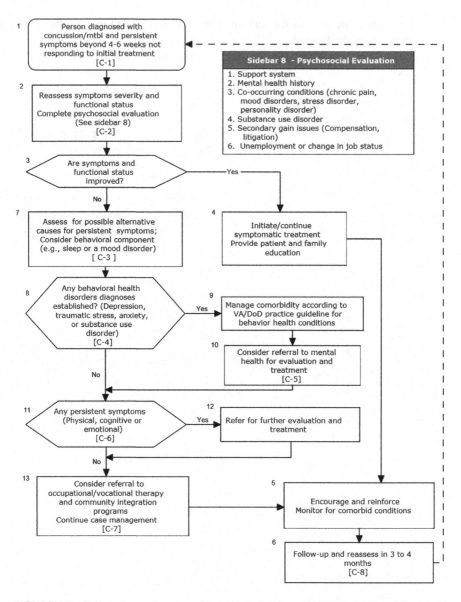

FIGURE 1.7   Follow-up persistent symptoms. Adapted from VA/DOD (2009, 2016).

Diffuse axonal injury occurs when axons are twisted, stretched, or severed (Brainline, 2020). This occurs due to acceleration/deceleration and contact forces producing a shearing effect. Nerve fibers (neuron body and axon) die and the brain swells, restricting blood supply and effecting brain herniation.

Patients often report lingering symptoms well after the injury event has occurred. These symptoms include dizziness, balance and coordination issues, fatigue, insomnia, and emotional or personality changes such as irritability, depression, and anxiety.

A complicating symptom is difficulty with information retrieval, since information about the injury event is needed to provide an accurate diagnosis.

While symptoms often decline rapidly, they may also last well beyond 3 months from the time of the event. They may be related to comorbidities (CDC 2020) including PTSD, major depressive disorder (MDD), suicidal tendencies, and substance abuse disorder (SUD). The cumulative effect is an increase in duration and severity of symptoms. Nonspecific symptoms associated with both mTBI and PTSD include irritability, insomnia, depression, fatigue, anxiety, and other cognitive defects. Such a mix of physical, genetic, and psychosocial factors may slow recovery times from weeks to years, as in the case of comorbid anxiety and depression.

BIOMECHANICS OF BRAIN INJURY

Cross-disciplinary research provides the insight necessary to fully understand the nervous system and its response to trauma. At the intersections between molecular neurobiology and biomedical engineering, for instance, is the field termed "molecular neuroengineering." According to Professor David Meaney's website at the University of Pennsylvania (Penn Engineering Directory, 2020):

> The process of mechanotransduction is critical in understanding the response of cells and tissues of the central nervous system to traumatic injury. In this research area, experimental work is combined with mathematical modeling to provide a method to quantify the effect of physical forces on cell and tissue function. For example, some of the research combines finite element models of the brain with experimental work to estimate the tissue mechanical stress/strain associated with biological markers of injury. …to determine the mechanisms by which a mechanical signal is converted into a biochemical signaling cascade.…

When a TBI occurs, axonal shearing, contusion, and coup/contrecoup injuries may occur. Axons can tear if sufficiently stretched. With substantial shearing, the injury is diffuse, throughout the brain, and cannot be treated. Contusions occur from bruising and bleeding due to a tear in small blood vessels. Coup injury occurs where the head is impacted and contrecoup occurs opposite the blow. Due to the irregular rough texture of the bones next to the frontal and temporal lobes, most coup/contrecoup injuries occur in these areas (Brain Injury Association of America, 2012).

Beneath our quarter-inch thick skulls are the meninges—three layers of connective tissue membranes that enclose the brain inside the skull. The three layers beneath the skull are the dura mater, arachnoid mater, and pia mater. If these layers are torn or infected, serious damage to the brain can occur.

HISTOLOGICAL EVIDENCE

McKee (2018) described the findings of a panel of neuropathologists in an initiative conducted by the National Institute of Neurological Disorders (NINDS) and the National Institute of Biomedical Imaging and Bioengineering (NIBIB) where the panel determined that

CTE has a pathognomonic lesion that distinguishes it from all other neurodegenerative diseases, including aging and nonspecific astrotauopathy (ARTAG). The pathognomonic lesion of CTE is defined as an accumulation of abnormal tau in neurons and astroglia distributed around small blood vessels at the depths of cortical sulci and in an irregular pattern. The panel also agreed that p-tau immunoreactive dot-like structures were characteristic of the pathology, including the perivascular lesions. Moreover, the TDP-43-immunoreactive inclusions in CTE were considered to be distinctive from other neurodegenerations, and the pattern of hippocampal degeneration was unlike the typical pattern found in Alzheimer's Disease.

FINITE-ELEMENT TECHNIQUES

The finite-element method uses numerical analysis to provide approximate solutions to partial differential equations (PDEs). In steady-state problems, the differential equations are completely eliminated; otherwise, the PDEs are transformed into an approximation consisting of ordinary differential equations and then, they are numerically integrated using techniques such as Euler's method, Runge-Kutta, and others (Zienkiewicz and Taylor, 1991).

Stress–strain vectors have been characterized in terms of acceleration/deceleration and rotation. Finite-element modeling indicates parts of the brain that are at a greater risk due to motion (McIntosh et al., 1996; Mao et al., 2010).

VA/DoD INJURY MECHANISM DESCRIPTIONS

Following are the injury mechanism descriptions resulting from the VA/DoD studies as part of the development of a clinical practice guideline (Rdaigan et al., 2018). The etiology provides a concise description of injury mechanisms and pathology.

In both blast and non-blast etiologies, primary injury can involve neurons, neuroglia, and vascular structures (Casey and McIntosh, 1994). A multitude of diffuse and dynamic processes also can contribute to secondary injury to include hypoxia and hypotension. The result of this secondary process is the release of inflammatory cytokines, initiation of an excitotoxic cascade, development of cerebral edema, and apoptotic signaling. The effects of free radical oxygen species, excitatory amino acids, and fluctuations in ion gradients such as calcium, alterations in neurotransmitters such as glutamate, receptor activation, lipid peroxidation, and mitochondrial uncoupling all result in increased neurologic injury. While the extent of such processes may be limited within the mTBI spectrum, the disturbances in brain metabolism and network connectivity associated with mTBI are related more to the complex cascade of ionic, metabolic and physiologic events rather than to structural injury or damage. The unique molecular activation and intracellular processes associated with individual mTBI etiologies require continued investigation. In addition, the effect of these physiologic responses needs to be studied over a variety of acute, sub-acute, and chronic time points in order to identify the underlying pathophysiology associated with mTBI and its association with the development of chronic neurodegenerative changes in a subpopulation of at-risk individuals.

An individual blast produces a complex mechanical profile consisting of a primary shock wave, followed immediately by a period of negative pressure, generation of a supersonic blast wind, and a delayed period of dissipating elevated pressure. However, depending on multiple blast and environmental variables this profile is quickly modified. Primary blast injury originates from early time point interactions between the blast-induced shock wave and the regional parenchyma and extra parenchymal tissues. This may result in a diffuse traumatic injury which precedes the onset of any linear or rotational acceleration injury. Passage of the shock wave through the tissues generates a combination of mechanical stresses which engage the neurons, glial cells, extracellular matrix, vascular structures, and cerebrospinal fluid-containing structures. These forces include spalling, shearing, mean and deviatoric stress, pressure, and volumetric tension. Secondary blast injury is related to objects which are displaced by the blast overpressure and blast wind. Secondary injuries may include a combination of both penetrating and blunt trauma. Tertiary blast injury occurs when an individual is thrown by the blast, sustaining blunt trauma such as a closed brain injury. Quaternary blast injuries, such as burns, chemical exposure, and asphyxia are directly related to the blast, but cannot be classified as a primary, secondary, or tertiary injury. Physical effects of the primary blast on an individual depend not only on blast characteristics but also on the physical relationship to the blast, such as the distance from the blast and exposure in an open environment versus an enclosed structure.

While isolated head trauma does occur, often mTBI blast-related mechanisms of TBI are associated with multisystem polytrauma and complicated by factors known to exacerbate secondary injury such as hypotension, hypoxia, and hypothermia, and primary blast effects on an individual likely do not often occur in the absence of secondary or tertiary blast effects, due to the narrower radius of primary blast dispersion compared with more widespread dispersion of blast fragment. The neuro-metabolic cascade following TBI is diverse and dynamic. The contribution of any particular physiologic response varies based on the magnitude of the forces involved, environmental features, and an individual's unique characteristics at the time of the event. Potential modifiers include, but are not limited to, genetic profile and epigenetic response to blast or non-blast stimulus, a history of previous TBI, general medical conditions, sleep deprivation, increased levels of stress hormone, and nutritional and hydration status.

Non-blast injuries are associated with focal, multifocal, and diffuse injury. Coup/contrecoup injury is the result of a mismatch in brain and skull movement. When the skull moves faster than the brain, the brain will strike the inner table of the calvarium causing a focal contusion, then, after the skull and brain have stopped their initial direction of movement, the brain may rebound in the opposite direction and impact the calvarium a second time. The orbitofrontal and anterior temporal lobes are most often affected as these are the most common sites of impact from motor vehicle accidents and sports-related injuries. The secondary effects of an acceleration/deceleration injury include edema and hemorrhage. Depending on the individual forces transmitted during an event, white matter injury through axonal stretch may play a prominent role in the pathology and clinical sequelae associated with both blast and non-blast mechanisms of TBI. With increased energy transfer, acceleration/deceleration is the primary etiology associated with diffuse axonal injury (DAI) and can occur as a primary mechanism of injury in closed brain injury or as a secondary force associated with blast exposure. A complex interrelationship exists between impact location, linear and rotational acceleration and concussion as a primary or secondary effect of acceleration/deceleration forces. To what extent the addition of shock wave propagation plays in modulation of biomechanical properties and what, if any, distinct physiologic effects are generated from the cumulative effects of blast plus acceleration, rather than either primary mechanism of injury in isolation, is currently unknown.

ANIMAL MODELS

Validated animal models for mTBI provide an opportunity to understand the mechanisms, pathology, and therapies. A range of closed head models appear in the literature and incorporate a variety of aspects of mTBI biomechanics (Bodnar et al., 2019). Model animals such as rodents are essential in TBI research to replicate the unique features of mTBI, including emotional and cognitive symptoms, biomechanics of impact and rotation, and to build a framework of common data elements for better-informed future research. Common test instrumentation such as controlled cortical impact (CCI), weight drop, and fluid percussion (FP) protocols assist in data normalization.

While promising neuroprotective drugs have been effective in animal TBI models, all have failed to produce effective results during Phase II or III clinical trials. The failure of preclinical study to translate into clinical implementation suggests the need to revisit TBI animal models and therapeutic strategies (Centers for Disease Control Prevention; National Center for Injury Prevention and Control; Division of Unintentional Injury Prevention, 2014). Areas requiring innovation include improved clinical trial design, optimization of therapeutic dosing and timing, determination of the effects of age, sex, and species or strains, and improvements in brain drug delivery (Xiong et al., 2013).

REFERENCES

AANS. 2020. *Neurosurgical conditions and treatments*, accessed March 28, 2020 at https://www.aans.org/Patients/Neurosurgical-Conditions-and-Treatments/Concussion.

Alexander, M. 1995. Mild traumatic brain injury: pathophysiology, natural history and clinical management. *Neurology*. 45:1253–1260.

American Psychiatric Association. 2013. *Diagnostic and statistical manual of mental disorders* (Revised 5th ed.). Arlington, VA: American Psychiatric Publishing.

Arrowhead Publishers. 2014. *Traumatic brain injury-therapeutic and diagnostic pipeline assessment and commercial prospects*. Chanhassen, MN: Arrowhead Publishers and Conferences.

Ashley, M.J. and D.A. Hovda, eds. 2018. *Traumatic brain injury-rehabilitation, treatment, and case management*. Boca Raton: CRC Press.

Assistant Secretary of Defense. 2015. *Traumatic brain injury: updated definition and reporting*. Washington, DC: Department of Defense.

Bear, M.F. and R.C. Malenka. 1994. Synaptic plasticity: LTP and LTD. *Curr Opin Neurobiol*. 4(3):389–399.

Blinzinger, K. and G. Kreutzberg. 1968. Displacement of synaptic terminals from regenerating motor neurons by microglial cells. *Z Zellforsch Mikrosk Anat*. 85:145–157.

Bodnar, C.N., K.N. Roberts, E.K. Higgins, and A.D. Bachstetter. 2019. A systematic review of closed head injury models of mild traumatic brain injury in mice and rats. *J Neurotrauma*. 36:1683–1706.

Brain Injury Association of America. 2012. *About brain injury*, accessed March 14, 2020. http://www.biausa.org/about-brain-injury.htm.

Brainline WETA-TV. 2020. Online course *Identifying and treating concussion/mTBI in service members and veterans*. Arlington, VA: Brainline, accessed February 3, 2020 at https://www.brainline.org/professionals/online-courses.

Brennan, P., A. McElhinny, and L. Mckinnon. 2014. The "Glasgow Coma Scale" an update after 40 years. *Nursing Times* 110:12–16.

Carney, N., J. Ghajar, A. Jagoda, et al. 2014. Executive summary of concussion guidelines step 1: systematic review of prevalent indicators. *Neurosurgery*. 75:S1–S2.

Casey, K.F. and T. McIntosh. 1994. The role of novel pharmacology in brain therapy. *J Head Trauma Rehabil*. 9(1):82–90.

Centers for Disease Control and Prevention; National Center for Injury Prevention and Control; Division of Unintentional Injury Prevention. 2014. *Report to congress on traumatic brain injury in the United States: epidemiology and rehabilitation*. Atlanta, GA: Centers for Disease Control and Prevention.

Centers for Disease Control. 2016. *Traumatic brain injury and concussion*, accessed February 11, 2020 at https://www.cdc.gov/traumaticbraininjury/index.html.

Centers for Disease Control. 2020. *Facts of physicians heads up facts for physicians about mild traumatic brain injury*, accessed February 12, 2020 at http://www.concussiontreatment.com/images/CDC_Facts_for_Physicians_booklet.pdf.

De Calignon, A., M. Polydoro, M. Suárez-Calvet, et al. 2012. Propagation of tau pathology in a model of early Alzheimer's disease. *Neuron*. 73:685–697.

Department of Defense. 2014. Traumatic Brain Injury: (Appendix G) from *Military Health System Coding Guidance: Professional Services and Specialty Medical Coding Guidelines (Version 4.0)*.

Farkas, O. and J.T. Povlishock. 2007. Cellular and subcellular change evoked by diffuse traumatic brain injury: a complex web of change extending far beyond focal damage. *Prog Brain Res*. 161:43–59.

Fortier, C.B., M.M. Amick, L. Grande, et al. 2014. The Boston assessment of traumatic brain injury-lifetime (BAT-L) semistructured interview: evidence of research utility and validity. *J Health Trauma Rehabil*. 29(1):89–98.

Fortier, C.B., M.A. Amick, A. Kenna, W.P. Milberg, and R.E. McGlinchey. 2015. Correspondence of the Boston assessment of traumatic brain injury-lifetime (BAT-L) clinical interview and the VA TBI screen. *J Head Trauma Rehabil*. 30(1):E1–E7.

Gentleman, S.M., P.D. Leclercq, L. Moyes, et al. 2004. Long-term intracerebral inflammatory response after traumatic brain injury. *Forensic Sci Int*. 146:97–104.

Gioia, G. A. and M.W. Collins. 2006. *Acute concussion evaluation (ACE)*, accessed March 28, 2020 at https://www.cdc.gov/concussion/headsup/pdf/ACE=a.pdf2006.

http://tbi-impact.org/, accessed March 10, 2020.

Joseph, R. 2011. *Head injury-skull fractures, concussions, contusions, hemorrhage, coma, brain injuries*. Cambridge: University Press.

Kandel, E. 2006. *In search of memory*. New York: W.W. Norton and Co.

Kay, T., D.E. Harrington, R. Adams, et al. 1993. Definition of mild traumatic brain injury. *J Head Trauma Rehabil*. 8:86–7.

Lifshitz, J. 2015. Experimental CNS trauma—a general overview of neurotrauma research. In Kobeissy, F.H., ed. *Brain neurotrauma-molecular, neuropsychological, and rehabilitation aspects*. Boca Raton: CRC Press.

LIMBIC-CENC website. *Long-term impact of military-relevant brain injury consortium chronic effects of neurotrauma consortium*, accessed March 31, 2020 at https://www.limbic-cenc.org/index.php/about/.

Mao, H., X. Jin, L. Zhang, et al. 2010. Finite element analysis of controlled cortical impact-induced cell loss. *J Neurotrauma*. 27:877–888.

Marshall, K.R., S.L. Holland, K.S. Meyer, E.M. Martin, M. Wilmore, and J.B. Grimes. 2012. Mild traumatic brain injury screening, diagnosis, and treatment. *Mil Med*. 177(8):67.

McCrory, P., W.H. Meeuwisse, M. Aubry, et al. 2013. Consensus statement on concussion in sport: the 4th international conference on concussion in sport. Held in Zurich. *Br J Sports Med*. 47:250–8.

McIntosh, T.K., D.H. Smith, D.F. Meaney, et al. 1996. Neuropathological sequelae of traumatic brain injury: relationship to neurochemical and biomechanical mechanisms. *Lab Invest.* 74:315–342.

McKee, A. 2018. Chronic traumatic encephalopathy. In Ashley, M.J. and D.A. Hovda, eds. *Traumatic brain injury-rehabilitation, treatment, and case management.* Boca Raton: CRC Press.

National Academies of Science, Engineering, and Medicine. 2019. *Evaluation of the disability determination process for traumatic brain injury in veterans.* Washington, DC: The National Academies Press. https://doi.org/10.17226/25317.

National Institute of Neurological Disorders and Stroke. 2012. The changing landscape of traumatic brain injury research. *The Lancet Neurology, Editorial, 11,* accessed February 10, 2020 at http://www.ninds.nih.gov/research/tbi/InTBIR_editorial_2012.pdf.

North, S.H., L.C. Shriver-Lake, C.R. Taitt, and F.C. Ligner. 2012. Rapid analytical methods for on-site triage for traumatic brain injury. *Annu Rev Anal Chem.* 5:35–56.

Omalu, B. 2008. *Play hard, die young, football dementia, depression, and death.* Lodi, CA: Neo-Forenxis Books.

Ontario Neurotrauma Foundation. 2013. *Guidelines for concussion/mTBI and persistent symptoms,* second edition. Toronto, Ontario: Ontario Neurotrauma Foundation.

Penn Engineering Directory, *David Meaney profile,* accessed March 14, 2020 at https://www.seas.upenn.edu/directory/profile.php?ID=64.

Radigan, L., R. McGlinchey, W.P. Milberg, and C.B. Fortier. 2018. Correspondence of the Boston assessment of traumatic brain injury-lifetime (BAT-L) and the VA comprehensive TBI evaluation (CTBIE). *J Head Trauma Rehabil.* 33(5):E51–E55.

Relias Academy (2020) is an online course entitled "The Fundamentals of Traumatic Brain Injury (TBI)" found at https://reliasacademy.com/rls/store/browse/productDetailSingleSku.jsp?productId=c529689.

Rosenfeld, J.V., A.I. Maas, P. Bragge, et al. 2012. Early management of severe traumatic brain injury. *Lancet.* 380:1088–1098.

Soto, C. 2012. In vivo spreading of tau pathology. *Neuron.* 73:621–623.

Teasdale, G., Maas, A., Lecky, F., Manley, G., Stocchetti, N., & Murray, G. (2014). The Glasgow Coma Scale at 40 years: Standing the test of time. The Lancet Neurology, 13(8), 844–854. https://doi.org/10.1016/S1474-4422(14)70120-6.

VA/DOD. 2016. *Clinical practice guideline for the management of concussion-mild traumatic brain injury version 2.0.* Department of Veterans Affairs and Department of Defense.

VA Research. 2019. *Assessment tool for military TBI.* Accessed April 4, 2020 at https://www.research.va.gov/research_in_action/Assessment-tool-for-military-TBI.cfm

Walker, W.C., W. Cane, L.M. Franke, et al. 2016. The chronic effects of neurotrauma consortium (CENC) multi-center observational study: description of study and characteristics of early participants. *Brain Injury.* 30:1469–1480.

World Health Organization. 1995. *Standards for surveillance of neurotrauma.* Geneva: WHO.

Xiong, Y., M. Asim, and M. Chopp. 2013. Animal models of traumatic brain injury. *Nat Rev Neurosci.* 14:128–142.

Zienkiewicz, O.C. and R.L. Taylor. 1991. *The finite element method,* vols. 1 and 2. New York: McGraw-Hill.

2 Sensing and Assessment of Brain Injury

FLUID-BASED BIOMARKERS

For the determination/detection of mTBI subsequent to an event such as an IED explosion, there is a wide range of evaluation methodologies one might consider. These include neurodiagnostic or psychological testing, neuroimaging modalities such as computed tomography or functional magnetic resonance imaging, balance assessment testing, EEGs, eye-tracking techniques, and genetic testing. But the requirement is to perform the test in a hostile environment, in a time frame suitable for therapeutic intervention. And we require both specificity and sensitivity for the tests to be performed.

Biomarkers serve to indicate normal or pathogenic processes, as well as pharmacological response to therapies. Prognostic biomarkers are indicative of patient survival and predictive biomarkers indicate therapeutic efficacies. Attributes of ideal biomarkers include the following (Arrowhead Publishers, 2014):

- reflect pathology of brain damage
- demonstrate high sensitivity and specificity for TBI
- correlate with neurological scores and neuroimaging data
- stratify patients by severity of injury
- appear in accessible biological fluid within minutes to hours subsequent to TBI
- provide information on injury mechanisms
- have well-defined biokinetic properties
- monitor progress of disease and response to treatment
- predict outcome of TBI patients
- easily measured by widely available simple-to-use platforms

Proteomic and systems biology approaches to biomarker discovery reduce the candidate markers to a manageable size for further analysis. The ultimate objective is to then develop a point-of-care handheld diagnostic device for use on the field, in theater, or in the emergency room. This would allow immediate triage of athletes and soldiers.

Perhaps the best metric for mTBI diagnosis in the field explored thus far is the use of biomarkers, as they are suitable for point of care diagnostics. Biomarkers are defined (Biomarkers Definitions Working Group, 2001) as those with a "characteristic that is objectively measured and evaluated as an indicator of normal biological processes, pathogenic processes, or pharmacologic responses to a therapeutic intervention." They may be implemented based on preliminary indications that mTBI may have occurred. Some key potential biomarkers follow along with their associations to a variety of cell damage mechanisms (Wang et al., 2015). S100B is a good example of a biomarker that exhibits not only in cerebral spinal fluid but can also be detected in blood.

MAP-2 dendritic marker, a neuron microtubule associated protein-regulating stability of tubulin subunits

MBP-demyelination marker, the myelin basic protein

S100B-brain injury biomarker, calcium-binding protein B marker for glial and Schwann cell damage

SBDP150/145-acute neural necrosis marker, alpha-spectrin cytoskeletal protein breakdown product

SBDP120-delayed neural apoptosis marker, alternate proteolysis in related axonal marker

UCH-L1-neuronal cell body injury marker, ubiquitin carboxyl-terminal hydrolase

SFAP-gliosis marker, glial fibrillary acidic intermediate filament protein specific to astrocytes

Also commonly associated with Alzheimer's disease, the genetic risk factor for CTE is the E4 allele of the apolipoprotein E (APOE) gene, which can dramatically impact the outcome of mTBI (Mannix and Meehan, 2015).

Certainly, a blood-based biomarker is more practicable than one requiring the extraction of spinal fluid, but this may require an extra step in which the blood-brain barrier is modulated to release sufficient quantities of markers—or panels of markers—for precise diagnosis. Figure 2.1 illustrates representative temporal variations among the neuronal pathologies and the biomarkers associated with cellular injury modalities. Relative concentrations in a biomarker panel may provide specificity related to the temporal neuronal pathology cascade.

Since we prefer minimally invasive validation and quantification, we desire a quantitative biomarker indicator of a specific biological or disease state that can be measured using samples taken from either the affected tissue or from peripheral body fluids. The markers should be available in detectable quantities in blood or ideally saliva rather than only cerebrospinal fluid (Wang et al., 2015). Challenges to proper identification stem from the minute quantities to be detected, along with the variable clearance rates. This necessitates detection technologies such as genomics,

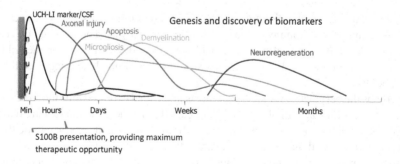

FIGURE 2.1 Temporal neuronal cascade following brain trauma.

neuroproteomics, and microRNA techniques. It is also desirable to distinguish aspects of PTSD and injury-induced psychological issues.

An excellent tool for analysis and interpretation of ohmics data for biomarker discovery is the Ingenuity® Pathway Analysis (IPA®) application (Qiagen Ingenuity, 2020). According to their website,

> IPA is a web-based software application for the analysis, integration, and interpretation of data derived from 'omics experiments, such as RNAseq, small RNAseq, microarrays including miRNA and SNP, metabolomics, proteomics, and small-scale experiments that generate gene and chemical lists. Powerful analysis and search tools uncover the significance of data and identify new targets or candidate biomarkers within the context of biological systems.

Genome-wide association studies of single nucleotide polymorphisms (SNPs) may be conducted through the National Human Genome Research database at the NIH (NIH GWAS, 2020) to catalog SNPs relevant to mTBI and to evaluate dynamic epigenetic effects associated with certain aspects of the brain injury (Perry, 2019) such as DNA methylation, histone modifications, and chromatin alterations. miRNAs differ by tissue type and will lead to the selection of candidate biomarkers for clinical trials (see also Appendix 3). Perhaps this approach can be further substantiated through the World Health Organization (WHO) international classification of function and the Precision Medicine Initiative launched by President Obama in his state of the union address in 2015 (Wagner, 2019).

Biomarkers may also be artifacts or imaging indicators related to injury or disease. They should be readily validated via fMRI or CT diagnostics and present within 24 hours subsequent to the trauma. Such theranostics will provide reliable prediction and efficacy of therapeutic opportunities. A practicable implementation would be suitable for POC field injury diagnostic instrumentation as well as relate directly to field testing metrics for helmets and body armor.

According to their company website (Banyon, 2018),

> Banyan Biomarkers has identified two protein biomarkers, Ubiquitin C-terminal Hydrolase-L1 (UCH-L1) and Glial Fibrillary Acidic Protein (GFAP), that are detectable in the blood shortly after TBI. UCH-L1 is primarily found in neurons and is involved in cellular protein regulation. GFAP is a member of the intermediate filament family of cytoskeletal proteins which form polymeric networks that provide structural support to glia, which support and nourish cells in the brain. Banyan Biomarkers believes that accurate diagnosis of TBI in acute care environments could significantly simplify decisions about patient management and improve medical care.

Banyan recently announced that the U.S. Food and Drug Administration (FDA) granted the company's request for the commercialization of Banyan BTI™ (Brain Trauma Indicator), a blood test utilizing Banyan biomarkers.

CHALLENGES IN BIOMARKER DEVELOPMENT

Biomarker research has advanced rapidly with development of reliable polypeptide ionization processes and bioinformatics. Neuronal dysfunction may be characterized via differential protein signatures, and hundreds of biomarker candidates for mTBI have been identified. The ultimate goal is to generate a validated, clinically relevant, and FDA-approved biomarker assay; but a range of technical, financial, legal, and regulatory milestones must be met before the system may be produced commercially (Ottens and Wang, 2009; Rifai et al., 2006; Vitzthum et al., 2005).

Proteins released as a result of brain injury into the cerebral spinal fluid occur in high concentration (Romeo et al., 2005) and enter the blood stream either through the blood-brain barrier or filtration of the CSF. A systems biology approach is used to generate candidate markers relevant to TBI pathobiology (Kobeissy et al., 2008). Concentration available and specificity of the marker to TBI determine the most suitable candidate markers. Pathologies mapped through the systems biology approach and some possible biomarker candidates are the following (after Kobeissy et al., 2008):

- Neuroinflammatory markers (gliosis: GFAP, S-100β; microgliosis: IL-6)
- Degradomic/cell death markers (calpain/necrosis, SBDP150/145: caspase-3/ apoptosis, SBDP120)
- Cell body injury/PTM markers (UCH-L1)
- Neuroregeneration markers
- Cytoskeleton damage markers (axonal, tau; dendritic, MAP2; demyelination, MBP)

Validation begins with preclinical animal models. Half-life and turnover may differ significantly between a rodent and a human patient and may lead to false assumptions in the assay qualification process (Ottens and Wang, 2009). Biomarker selection relies on the availability of a sensitive and reliable readout—usually an enzyme-linked immunosorbent assay-based test (ELISA) employing antibodies. Such epitope-specific binding molecules are sometimes available in commercial kits (e.g., https://www.biocompare.com/) but production of a specific antibody is usually required.

It is critical for the biomarker to be reliably detectable within a clinically relevant dynamic range. Sensitivity of ELISA assays is often improved with tyramide signal amplification (TSA), chemiluminescence detection, and immune-PCR (Ottens and Wang, 2009). Reliability and reproducibility of a bioanalytical method are demonstrated through a number of parameters including (Ottens and Wang, 2009) accuracy, precision, selectivity, sensitivity, reproducibility, and stability. Biomarker validation must demonstrate that the marker supports TBI diagnosis with clinical significance. "Analytic validation requires the biomarker to be measured in a test system with well-established performance characteristics. Clinical validation requires the biomarkers function within an established body of evidence that elucidates the physiologic, pharmacologic, and clinical significance of the test results" (Ottens and Wang, 2009).

The Commissioner of the U.S. Food and Drug Administration, Andrew von Eschenbach, MD, referred to biomarkers as one technology "most likely to modernize and transform the development and use of medicines." Such growing recognition resulted in the 2006 formation of the Biomarkers Consortium, a partnership including the NIH, FDA, Centers for Medicare and Medicade Services, and other industry representatives (Kobeissy et al., 2008).

SENSORS FOR mTBI

There is a need for sensors that allow laboratory or field detection of an event, for example an event created by a blunt force or a blast force. Such events are commonly found to be the cause of traumatic injuries such as traumatic brain injuries. An "event force" is any force type suitable to produce or model a traumatic brain injury of any form. Such forces include, but are not limited to, blunt force, ballistic force, and shock wave forces associated with blast trauma. With traumatic brain injuries, and particularly with mild traumatic brain injuries, there may be no external signs of injury which potentially could delay treatment, or give an indication that no treatment is necessary, leading to severe and often cumulative consequences (Mentzer, 2011).

A significant technology gap exists in the testing of personal protective equipment for subject individuals and animals, relating to body armor as well as helmet systems, and other protective equipment. Sensors are required to determine the correlation between threats (insults) to the subject so as to provide a means by which protective equipment can be assessed for its ability to protect a subject from a variety of insults and injury, and to optimize the design trade-offs between armor weight, thickness, energy dissipation, stopping power, and the like. This need extends to the widely publicized concerns regarding protection of subjects in conflict or competitive areas, and to the protective measures needed for contact sports such as American football (Mentzer, 2013).

Clearly a means is required by which the range of insults, including blunt trauma, ballistic impact, and shock trauma, can be measured with a metric that directly indicates the injury to the body due to an insult thereby directly correlating insult to injury, whereas a host of sensors have been employed to this end—including pressure sensors, accelerometers, strain sensors, and optical surface measurement methodologies to characterize the energy impacting the protective armor, and the dissipation of that energy through human tissues and a range of torso and head form anthropomorphic test modules (ATM) incorporating these sensors—the point and 2-dimensional energy characterizations, along with time-resolved networked sensor determinations, have provided a less than satisfactory correlation of insult to injury. While a host of candidate sensors continue to emerge in the literature (hydro gels, functionalized nanoparticles, photonic crystals, etc.), only the novel sensor concept disclosed herein (see Appendix 1) directly represents the response of human tissues to traumatic insult. Nanotechnology research is replete with examples of self-assembled chemicals forming well-controlled supramolecular films and structures, including manipulation of material properties at the atomic level of detail.

Problems with current sensors used in test labs include lack of repeatable measurement, poor to no correlation, lack of calibrated response to the range of insults to include ballistic threats; and concurrently, lack of correlation to any or all of the range of tissue susceptibilities and widely varying vulnerabilities. Test artifacts abound due to a wide range of variables, including threat mass, velocity, total yaw at impact, yaw cycle precession, obliquity at impact, backing material variability, along with backing material inconsistencies, tissue simulant variation, and lack of controlled test protocols proving repeatability of test metrics. This results in highly conservative limits for penetration depth at prescribed impact kinetic energies, providing only partially correlated determination of armor suitability—and little trade space for the armor designer to effect improvements (Mentzer, 2015).

Current methodologies for biomechanical and biomedical testing exhibit a bifurcation of experiments and data sets due to lack of sensor elements that relate the various data sets. Consider the desire to characterize an explosion event with a metric that provides the quantitative basis for incremental improvements in armor design, and for testing one armor design against another. We also desire a measurement that provides prediction of injury to a Soldier as a function of the event characterization. Further, we'd like to know, based on a particular event metric, not just the injury sustained, but what therapeutic opportunities exist as a function of time, based on the dataset produced by this same metric. Figure 2.2 illustrates the biomechanical/biomedical divide.

Blunt and ballistic traumas may exhibit the secondary effect of shock waves that travel through the material medium; and proper sensing relationships to injury modeling must account for the convoluted effect of all such insults to the PPE that are transmitted to the soldier. Propagation of blast waves is very complex, involving factors of force, acceleration peak pressure, duration, and strain. It could involve both direct propagation through the skull and indirect propagation via blood vessels—as well as neuronal scaffold tissue, such as actin and the integrin proteins binding nerve cells to the neuronal scaffold. The maximum blast effect is on "hollow" organs such as lungs, eardrums, and intestines. Because of the complexity of an explosive event scenario in which blunt traumas may be superimposed with shock waves and other secondary injury effects, the physical parameters currently used to characterize the events do not correlate well to injury.

A sensor metric or correlation test factor (CTF) is needed to bridge the bifurcation of biomechanical and biomedical data bases. It should be usable in any medium such as animal, cadaver, or other human surrogate experiments as necessary, be compatible with biomarker platforms, and calibrated for blunt, blast, and ballistic events causing mTBI. While current approaches focus on pressure, acceleration, peak

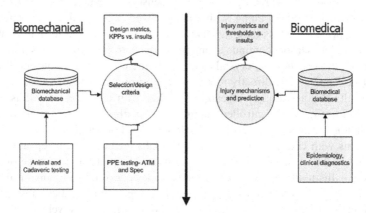

FIGURE 2.2 Bifurcation of biomechanical/biomedical knowledge bases. M.A. Mentzer, 2013 "Developing a Metric to Predict Mild Traumatic Brain Injury," presentation to National Academy of Science.

pressure, and force or strain, with unreliable correlation of insult to injury—the CTF should directly relate equipment design to injury. This would lead to new insights for PPE and mTBI injuries, improve predictive models, and enhance field diagnosis.

One such CTF is the device (Mentzer, 2012) described in Figure 2.3. This sensor employs self-assembled liposomes encapsulated in solution. The embodiment shown starts with dipalmitoylphosphatidylcholine (DPPC)—a common lipid found in brain tissues, and representative of the protective membranes surrounding nerve cells. Disruption of the liposomes encapsulating self-quenching calcein dyes produces a colorimetric change proportional to the traumatic event.

Disruption of the bilayer molecules in the liposome sensor might also be measured by a change in chirality of the three-dimensional molecular interaction between lipid layers—as well as change in chirality due to symmetry breaking in nonchiral molecular orientation—providing quantification and validation of the method. Correlation to biomechanical testing and validation as an objective CTF metric could be achieved with fluid percussive models and test apparatus shown in Figure 2.4.

This approach to the CTF will tie directly to the wealth of existing medical data by replicating conditions for fluid percussion models using the fluid percussion system built for these experiments. Brief fluid percussion pulses are presented to encapsulated liposome solutions in accordance with published injury parameters in the literature, and biomarker production and biochemical, molecular, and related

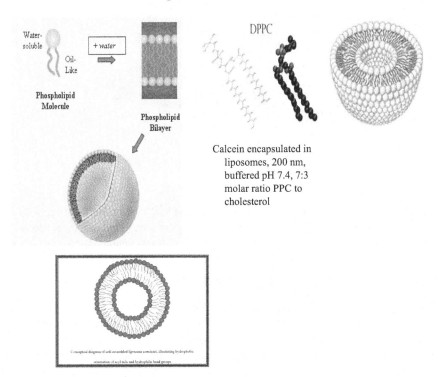

FIGURE 2.3   Liposome test structures. Courtesy of Dr. Zahra MirAfzali, Encapsula NanoSciences.

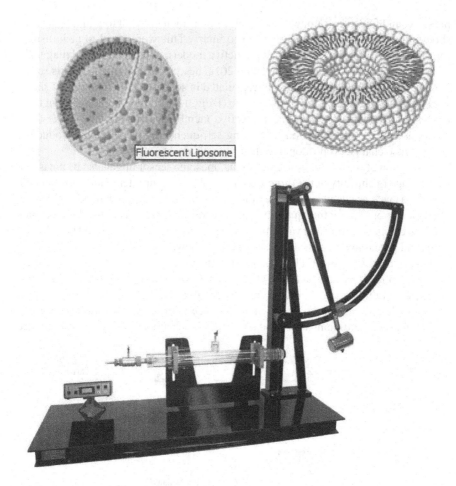

Fluorescent Liposome

FIGURE 2.4   Validation of CTF metric using fluid percussion device.

processes studied as a function of trauma presented to the brain. Replication of these conditions serves as a basis for not only calibrating the sensor, but as a direct correlation of biomarkers one would predict for identical trauma events.

ASSESSING FUNCTION WITH IMAGING

A variety of imaging techniques are used to characterize brain function, determine extent of an injury, and screen for tumors, stroke, or ischemic damage. The primary imaging modality for mTBI is magnetic resonance imaging (MRI). This provides the means to check for blood, brain contusion, and any significant damage to axons that result in lesions.

MR profusion or arterial spin labeling images blood flow to the brain regions. While not an imaging technique, MR spectroscopy examines brain tissue biochemical metabolites, providing a measure of neuronal and axonal integrity—thereby assessing the amount of pathology that has occurred (Jordan, 2012). But most brain injuries produce only normal CTs and MRIs. Diffusion tensor imaging (DTI) has emerged as a research tool to study mTBI. Neuropsychological evaluations also provide indications of extent of damage and functional effects.

A further comparison of brain injury imaging technologies (Brainline, 2020) is as follows:

PET-positron emission tomography-measures brain metabolism and indicates oxygen and glucose usage

T1-weighted MRI—a standard measure of anatomy and structure

T2-weighted MRI—for visualization of severe diffuse axonal injury

DWI-diffusion weighted MRI shows alterations in tissue integrity

SWI-susceptibility weighted imaging shows microhemorrhages and vascular insult

FLAIR MRI-fluid-attenuated inversion recovery—like DWI and sensitive to water content in brain tissue

DTI shows white matter tracts and brain connectivity; still a research tool

GRE-gradient record MRI shows hemorrhaging and blood in brain tissue; better than CT scans to detect microbleeds

fMRI looks at local increase in iron when neurons fire during various tasks; still a research tool

CT scans are of limited use in assessment of mTBI and will usually report negative results in mTBI patients (Stiell et al., 1997), as CT only detects structural brain damage. Small hemorrhages may also go undetected, especially with diffuse mTBI (Arrowhead Publishers, 2014). Brain swelling is measured using intracranial pressure (ICP) monitoring.

PET and fMRI provide understanding of brain biochemistry but DTI is needed to image structural changes like diffuse axonal injury (Carroll and Rosner, 2011). Successful TBI management depends critically on proper diagnosis. CTI will confirm intracranial hemorrhage, hematomas, and contusions associated with TBI, while mTBI is more difficult to diagnose (Arrowhead Publishers, 2014).

ELECTROPHYSIOLOGICAL ASSESSMENT OF BRAIN FUNCTION

Electro-chemical brain activity generates electrical potentials transmitted through neurons. The electrical signatures known as brain waves may be recorded to monitor brain function. Multi-channel electrodes placed on the scalp reflect cortical activity. Digitized recordings may be processed using mathematical algorithms for statistical analysis and brain mapping. Event-related potentials are extracted from electro-encephalograms (EEGs) to provide indication of the brain's response to a stimulus (Arrowhead Publishers, 2014).

Quantitative analysis of EEGs (qEEGs) examines frequency and amplitude characteristics to correlate EEGs to features of mild traumatic brain injury, but interpretation of results remains controversial, since patient variations relate to many factors (Haneef et al., 2013). Magnetoencephalography (MEG) experiments that measure the magnetic fields produced by electrical brain current indicate high-frequency gamma brain waves and potential relationships to brain injury in the pre-frontal and posterior parietal lobes of the cerebral cortex. The technique has been proposed as a measuring tool to diagnose the efficacy of brain stimulation therapies such as transcranial electrical stimulation (TES) and transcranial magnetic stimulation (TMS) (Huang, 2019).

Neurons transmit information interneuronally via chemical signals, where the Na/K-ATPase establishes and maintains [Na] and [K] gradients across the cell membrane, and interneuronally via electric signals. The flow of inorganic ions across the cell membrane is responsible for the generation of electrical signals inside and outside the neuron. Frequency and time domain analysis of these signals, temporal response dynamics, and the clocking mechanisms involved in brain function relates structure to function in a complex manner (Buzsaki, 2006). Additional research in the analysis of brain waves may provide useful methodologies for diagnosis and treatment of mTBI.

U.S. FOOD AND DRUG ADMINISTRATION REGULATION OF NEUROLOGICAL DEVICES

The Center for Devices and Radiological Health (CDRH) of the Food and Drug Administration (FDA) regulates the review and approval process for neurological devices. The FDA research includes (FDA website, 2020):

- The development of more reliable neural interface devices.
- Identification of new systems-level biomarkers for neurological diseases.
- Comparisons of invasive and noninvasive neural interface devices to characterize both safety and efficacy over long periods of time.
- New methods of analysis and data visualization to extract as much useful information as possible from physiological recordings.
- Developing test methods for upper limb performance and cognitive load to evaluate upper limb prosthetics.
- Changes in neurophysiology, anatomy, behavior, and electrode material properties caused by cortical implants.
- Performance metrics to assess the long-term safety and performance of peripheral nerve interfaces.
- Electronic modeling techniques that can be used for magnetic resonance imaging (MRI) testing.
- Detection of brain injury.
- Applications of high intensity therapeutic ultrasound to diagnosing and treating brain injuries.
- Long-term performance of neural implants.

Medical devices currently approved for assessment of head injury are listed on the FDA website, along with a premarket database for each device:

Manufacturer	Device
Banyan BTI*	Brain Trauma Assessment Kit
BrainScope Company, Inc.	BrainScope Ahead 100
	Ahead 200
	Ahead 300
	BrainScope One
	Modified BrainScope One
	BrainScope TBI (Model: Ahead 400)
ImPACT Applications, Inc	ImPACT
	ImPACT Quick Test
	ImPACT
InfraScan	Infrascanner Model 1000
	Infrascanner
Oculogica, Inc.	EyeBOX

Approvals assess the intent to diagnose, intended population, and a locked-down assay for introduction into the U.S. market. The FDA receives thousands of pre-submission packages per year (Jeffrey, 2019). Unfortunately, many interesting experiments do not translate from mouse to man.

REFERENCES

Arrowhead Publishers. 2014. *Traumatic brain injury—therapeutic and diagnostic pipeline assessment and commercial prospects.* Chanhassen, MN: Arrowhead Publishers and Conferences.

Banyon. 2018. *Biomarkers*, accessed February 24, 2020 at http://www.banyanbio.com.

Biomarkers Definitions Working Group. 2001. Biomarkers and surrogate endpoints: preferred definitions and conceptual framework. *Clin Pharmacol Ther.* 69:89–95.

Brainline WETA-TV. 2020. Online course. *Identifying and treating concussion/mTBI in service members and veterans.* Arlington, VA: Brainline, accessed February 3, 2020 at https://www.brainline.org/professionals/online-courses.

Buzsaki, G. 2006. *Rhythms of the brain.* New York: Oxford University Press.

Carroll, L. and D. Rosner. 2011. *The concussion crisis anatomy of a silent epidemic.* New York: Simon and Schuster.

FDA website. 2020. *Regulatory science for neurological devices*, accessed April 1, 2020 at https://www.fda.gov/medical-devices/neurological-devices/regulatory-science-neurological-devices.

Haneef, Z., H. S. Levin, J.D. Frost, and E.M. Mizrahi. 2013. Electroencephalography and quantitative electroencephalography in mild traumatic brain injury. *J Neurotrauma.* 30(8):653–656.

Huang, M. 2019. Marked increases in resting-state MEG gamma-band activity in combat-related mild traumatic brain injury. *Cerebral Cortex.* doi:10.1093/cercor/bhz087.

Jeffrey, D. 2019. *Regulatory considerations for blood-based biomarkers of TBI.* Pittsburgh: National Neurotrauma Society: Neurotrauma.

Jordan, B.D. 2012. Repetitive head injury in athletes. *Second Annual Johns Hopkins Traumatic Brain Injury National Conference on Repetitive Head Injury.* Baltimore.

Kobeissy, F.H., S.F. Larner, S. Sadasivan, et al. 2008. Neuroproteomic and systems biology-based discovery of protein biomarkers for traumatic brain injury and clinical validation (review). *Proteomics Clin Appl*. 2:1467–1483.

Mannix, R. and W.P. Meehan. 2015. Evaluating the effects of APOE4 after mild traumatic brain injury in experimental models. In Kobeissy, F.H., ed. *Brain neurotrauma—molecular, neuropsychological, and rehabilitation aspects*. Boca Raton: CRC Press.

Mentzer, M.A. 2011. *Applied optics, fundamentals and device applications—nano, MOEMS, and biotechnology*. New York: CRC Press Taylor and Francis Group.

Mentzer, M.A. 2012. *Blast, ballistic and blunt trauma sensor exhibiting differential circular dichroism chirality shifts and color changes based on concurrent disruption of tissue and sensor phospholipid bilayers configured as liposomes*. Provisional Patent 9080984, Alexandria: US Patent and Trademark Office.

Mentzer, M.A. 2013. *Analysis and design of a photonic biosensor for mild traumatic brain injury*. Aberdeen: Army Research Laboratory.

Mentzer, M. A. 2015. *Blast, ballistic and blunt trauma sensor*. Patent 9080984, Alexandria: United States Patent Office.

NIH GWAS. 2020. *Genome-wide association studies fact sheet*, accessed March 3, 2020 at https://www.genome.gov/about-genomics/fact-sheets/Genome-Wide-Association-Studies-Fact-Sheet.

Ottens, A. K. and K.K.W. Wang, eds. 2009. *Neuroproteomics methods and protocols*. New York: Humana Press.

Perry, Y. 2019. *Genetic markers of TBI outcome*. Pittsburgh: National Neurotrauma Society: Neurotrauma.

Qiagen Ingenuity website. 2020. Accessed March 3, 2020 at http://pages.ingenuity.com/rs/ingenuity/images/IPA_data_sheet.pdf.

Rifai, N., M.A. Gillette, and S.A. Carr. 2006. Protein biomarker discovery and validation: the long and uncertain path to clinical utility. *Nat Biotechnol*. 24:971–983.

Romeo, M.J., V. Espina, M. Lowenthal, B.H. Espina, E.F.I. Petricoin, and L.A. Liotta. 2005. CSF proteome: a protein repository for potential biomarker identification. *Expert Rev Proteomics*. 2:57–70.

Stiell, I.G., G.A. Wells, K. Vandemheen, et al. 1997. Variation in ED use of computerized tomography for patients with minor head injury. *Ann Energ Med*. 30(1):14–22.

Vitzthum, F., F. Behrens, N.L. Anderson, and J.H. Shaw. 2005. Proteomics: from basic research to diagnostic application. A review of requirements and needs. *J Proteome Res*. 4:1086–1097.

Wagner, A. 2019. *Rehabilomics research model as a framework for precision rehabilitation*. Pittsburgh: National Neurotrauma Society: Neurotrauma.

Wang, K.W., Z. Zhang, and F.H. Kobeissy, eds. 2015. *Biomarkers of brain injury and neurological disorders*. Boca Raton: CRC Press Taylor & Francis Group.

3 Instrumentation for Assessing mTBI Events

PARAMETRIC ANALYSIS OF VIDEO

Video data analysis of a potential brain injury event requires proper camera setup, timing, triggering, calibration, data acquisition, storage and transfer, and ultimate parametric analysis and reporting. This is accomplished in a seamless workflow with optimal efficiencies, ensuring digital image data is optimized for parametric analysis and event characterization. Ancillary technologies include advanced illumination techniques, such as high brightness laser illumination, arc discharge lamp illumination, plasma lighting, structured lighting, and signal-enhancement techniques (Mentzer, 2011). Custom MATLAB routines are often employed for analysis of X-ray imagery. Specialized image acquisition methodologies are enhanced with data fusion from fiber-optic sensors and strain gauges, providing enhanced event characterization (Medina et al., 2012).

A typical process flow is illustrated in Figure 3.1 for an event employing, for example, high speed video acquisition, X-ray cineradiography, surface deformation, and an integrated sensor suite. Event characterization may include analysis of impact velocity and velocity curves, pitch and yaw during trajectory and at impact, and deformation, providing correlation to a range of parameters.

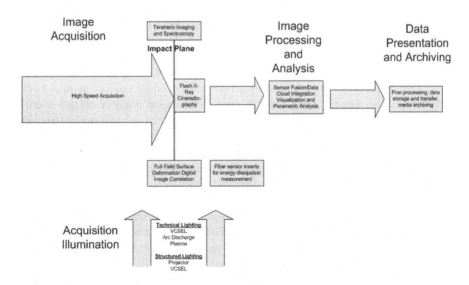

FIGURE 3.1 Characterization of an event related to mTBI. M.A. Mentzer, 2013 "Developing a Metric to Predict Mild Traumatic Brain Injury," presentation to National Academy of Science.

A typical problem uses an image as an intermediate data structure, formed from raw data, which is analyzed further to extract desired information. This information—such as the location of a defect in a body armor image—can best be determined only after thoroughly understanding the physics of the imaging process and characterizing all available prior knowledge. Given this understanding, we develop signal-processing algorithms that involve concepts from random signals, statistical inference, optimization theory, and digital signal processing (Prince and Links, 2006). A host of high-speed cameras are available commercially, including Vision Research Phantom and Photron high-speed cameras.

Analysis of Digital Imagery

Digital image processing suites available for commercial licensing include Aramis, TrackEye, ProAnalyst, Correlated Solutions, PixelRay, and developmental MATLAB routines. Customized routines for specific test applications are available, along with data fusion processes for enhanced visualization and accuracy.

2D Parametric analysis requires a single imager with a calibration sample in the plane of motion either before or during the event. This provides measurement of 2D flight characteristics such as pitch, yaw, position, velocity, and acceleration as well as angle, distance, and area, using the interactive feature tracking overlays. Distance and velocity, e.g., gun bolt movement or the height an object moves into the air from a mine blast, are readily determined from 2D imagery.

3D Parametric analysis requires at least two imagers and a 3D calibration fixture. This provides 3D flight characteristics such as pitch, yaw, position, velocity, and acceleration as well as angle, distance, and area. Coordinate location and distance calculation for a traveling object are derived from the world reference points collected in a 3D calibration. This is compared with time or fused with data from additional sensors.

Pitch and yaw for slugs and projectiles traversing a known path are calculated with a single 2D image or, with two cameras, a 3D image of the projectile flight. Area and volume for smoke cloud or flash evaluation are performed in 2D and utilize a 2D calibration in the event plane. Volume measurements utilize 3D analysis and calibration.

Surface deformation/3D image correlation can be used for measurement of back face deformation during armor tests. Surface deformation analysis utilizes a uniform random black and white speckle pattern on the surface and two calibrated high speed cameras. Full field deformation analysis provides displacement and velocity in three coordinates, x–y strain, principal strain angle, in-plane rotation, 3D contours, and curvature angles. Complementary solutions include projected structured lighting, laser illumination/bandpass filtering, and specialized illumination techniques (Mentzer, 2011).

An excellent example of optical 3D measurement of surface deformation is the ARAMIS system produced by the Gesellschaft für Optische Messtechnik, GOM mbh. Following is an excerpt from the GOM literature, illustrating the range of optical measurement capability for the 3D software suite.

ARAMIS helps to better understand material and component behavior and is ideally suited to monitor experiments with high temporal and local resolution.

ARAMIS is a noncontact and material-independent measuring system providing, for static or dynamically loaded test objects, accurate:

- 3D surface coordinates
- 3D displacements and velocities
- Surface strain values (major and minor strain, thickness reduction)
- Strain rates

ARAMIS facilitates:

- Determination of material properties (R- and N-values, FLC, Young's modulus)
- Component analysis (crash tests, vibration analysis, durability studies)
- Verification of finite-element analysis

ARAMIS is the unique solution delivering complete 3D surface, displacement, and strain results where a large number of traditional measuring devices are required (strain gauges, LVDTs extensometers).

The same system setup is used for multiple applications and can be easily integrated in existing testing environments.

REAL-TIME DATA PROCESSING AND INTEGRATION

ARAMIS provides an import interface for CAD data, which are used for 3D coordinate transformations and 3D shape deviation calculations. The import interface handles the following formats:

- Native: Catia v4/v5, UG, ProE
- General: IGES, STL, VDA, STEP

ARAMIS software provides real-time results for multiple measurement positions from the test objects surface. These are directly transferred to testing devices, data acquisition units, or processing software (e.g. LabView, DIAdem, MSExcel) and are used for:

- Controlling of testing devices
- Long-term tests with smallest storage requirements
- Vibration analysis
- 3D video extensometer

VERIFICATION OF FINITE-ELEMENT SIMULATIONS

As part of complex process chains, optical measuring systems have become important tools in industrial processes. Together with numerical simulation they have significant potential for quality improvement and optimization of development time for products and production. ARAMIS supports full-field verification of FE simulations. Determining material parameters helps in the evaluation and improvement of existing material models. Import of FE result datasets allows numerical full-field comparisons with FE simulations for all kinds of component tests. Thus, finite-element simulations can be optimized for reliability.

X-RAY IMAGING

X-ray imaging applications require unique optical detector system configurations for optimization of image quality, resolution, and contrast ratio. X-ray photons from multiple anode sources create a series of repetitive images on fast-decay scintillator screens, from which an intensified image is collected as a cineradiographic video (or intensified image) with high-speed videography. Current developments address scintillator material formulation, flash X-ray implementation, image intensification, and high-speed video processing and display. Determination of optimal scintillator absorption, X-ray energy and dose relationships, contrast ratio determination, and test result interpretation are necessary for system optimization (Mentzer, 2011).

Numerous test requirements motivate the development of flash X-ray cineradiography systems with multi-anode configuration for repetitive imaging in closely spaced time frames (Mentzer et al., 2010). Applications include the following:

- Shaped charge detonations to further understand properties of jet formation and particulation
- Explosively formed projectile detonations to quantify launch and flight performance characteristics
- Detonations of small caliber grenades and explosive projectiles to verify fuse function times and fragmentation patterns
- Performance and behavior of various projectiles and explosive threats against passive, reactive, and active target systems
- Human effects studies including body armor, helmets, and footwear
- Behind armor debris studies of large caliber ammunition against various armor materials
- Small caliber projectile firings to study launch, free flight, and target impact results

Specific parameters of interest include impact velocity of projectiles striking various materials, pitch, yaw, and roll during projectile trajectory and at impact, time to maximum deformation of materials, time to final relaxation state, and profile of the deformation occurring during the events of interest.

FLASH X-RAY

Flash X-ray is a well-established technology in which an electron beam diode (anode and cathode) in a sealed tube produces a brief (<30 ns) intense X-ray pulse. Recovery times are too slow to repetitively pulse the anode for high speed cineradiography—high frame speed requires multiple independent pulsers and diodes. Sources are closely packed to minimize parallax problems and to maintain source intensity with smaller diodes (Mattsson, 2007). Triggering of X-ray pulses is tailored to the event timing, so that precise imaging of the event is achievable with either high-speed framing cameras in which image frames are matched to the X-ray pulses, or high-speed video cameras with suitable exposure rates. Image processing algorithms are then utilized to allow extraction of parametric data from multiple frame X-ray images produced on scintillator screens and images through microchannel plate/image-intensifier hybrids.

X-ray computed tomography (XCT) represents an alternative approach with the advantage of a 3D image. The technique requires four or more pulsers for each 3D frame, translating to 32 or more pulsers to produce an equivalent of eight frames (Karsten and Helberg, 2005). There is no commercially available anode arrangement that lends itself to the correct geometry, and the cost of such a system would be substantial. For this reason, custom systems are constructed for specific applications (Mentzer, 2011).

TYPICAL SYSTEM REQUIREMENTS

System specifications for cineradiography systems in development include the following:

- multi-anode X-ray system, 150–450 keV X-ray photons
- fluorescent screen/real-time video camera imaging; 6–8 foot standoff
- 1–5 mm spot size, with $1\,m^2$ target area
- Up to 4 ms record time
- ≥ 8 images at 100,000 frames per second; trigger synch pulses
- pulser repetition rate adjustable from 10,000/s to 100,000/s
- independent inter-pulse time from anode to anode
- adaptable to a wide range of test scenarios/operation under harsh conditions/explosive test environment

Shown in Figure 3.2 is a multi-anode arrangement where the acquisition is achieved with a fast scintillator and a high-speed video camera. Multiple frames of imagery are produced on a scintillator screen by flash X-ray exposure through the test materials at either of 150 or 450 keV. A turning mirror is used to avoid placement of the video camera directly in the residual X-ray beam. In a former approach, separate X-ray heads were used to image the event from different directions without overlap of the projection onto storage phosphor screens or X-ray films, from which still video images were obtained. This approach creates parallax and geometry issues, however, motivating the current approach in which successive frames are captured on a single area location on a scintillator screen. A framing camera with an image intensifier is synchronized with the X-ray pulses to capture snapshots over events of duration from tens of microseconds to a few milliseconds (Mentzer, 2011).

The choice between 150 and 450 keV X-rays represents a tradeoff between material penetration and contrast ratio. The attenuation coefficient for clay, for example,

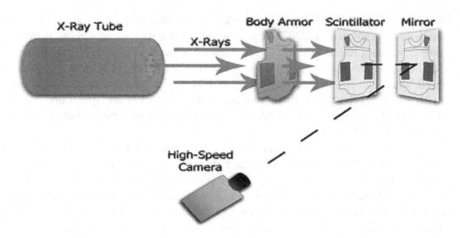

FIGURE 3.2 Typical multi-anode system configuration.

as reported by NIST (NIST, 2010), is $0.2\,cm^{-1}$ at 450 keV and $0.3\,cm^{-1}$ at 150 keV. For 10 in. of clay, the 150 keV flux is attenuated by about 15× with respect to a 450 keV pulse. It is possible, using 450 keV X-rays and a storage phosphor screen, to image a 0.5 in. depression under the described angular alignment. The attenuation differential with X-ray energy implies a contrast ratio difference. As it is desirable to image depressions smaller than 0.5 in. (1.27 cm), with optimal resolution, the improved contrast ratio that obtains for 150 keV X-rays is an advantage—if the higher flux requirement can be accommodated.

Scintillators for Flash X-Ray

Six commercially available fast scintillators, from two manufacturers and one distributor, were characterized for efficiency and speed at 150 and 450 keV, as efficiency differences impact the increased flux requirements for imaging at 150 keV. No information on uniformity or sensitivity was available, and the speed of the samples deviated from data published in the literature. Availability, pricing, and timely delivery of suitable scintillators is expected to be a challenge associated with the assembly of multiple systems (Mentzer, 2011).

The scintillators had various sizes up to about six inches, so they were masked to have a common 6-mm diameter circular area from which light could be collected. Each was placed in a dark chamber with a fast photomultiplier tube module (Hamamatsu H9656MOD), containing a transimpedance amplifier. X-rays from the pulsers were directed at the scintillators from about 8 in. outside of the dark box, at a flux density to the order of $10^9/cm^2$ at approximately 30 ns/pulse. The photomultiplier was located about 14 in. from the scintillator, and was followed by a Tektronix TDS 2024B 200 MHz, 2 Gs/s oscilloscope. Photographs of the resulting traces were used to report the integrated output and speed of the scintillators (Mentzer, 2011).

It is interesting to note that the efficiency at 150 keV is not very different from that at 450 keV, despite the 7.4× higher attenuation reported in the literature for 150 keV with respect to 511 keV (National Institute of Standards and Technology Physics Laboratory, 2010). Future work around the visible photon yield per X-ray, as a function of -ray energy and scintillator flavor, may help clarify the issue. It is also notable that the measured decay times are far longer than those reported (Shionoya and Yen, 1999). For frames of about 1 µs duration, placed at various spacing throughout an event of tens to thousands of microseconds, the scintillator response times are inadequate.

System Configurations

Using currently available technology, the apparatus typically required in order to implement a flash cinematography system will likely include:

- High-voltage power supply
- Trigger, timing, and delay control electronics
- Trigger amplifier
- X-ray Pulser(s) + [1] multi-anode X-ray tube
- Target positioning and backing fixture
- High decay rate scintillating screen (<10 µs decay)
- Mirror
- Image intensifier
- High frame rate camera
- Computer control system and software
- Connecting cables

This equipment can be configured to provide an efficient and safe means of capturing X-ray image sequences synchronized to high-velocity test events. The use of a multi-anode tube for X-ray exposure of fixed position targets will reduce the complexity of some test arrangements. Complete characterization of any parallax effects from multiple anode spacing can be used to enhance spatial resolution of images used for metrology. The imaging system selected for such a test configuration will require careful consideration of scintillating screen fluorescence, decay characteristics, and resolution limits. Similarly, the image intensifier resolution characteristics should be well matched to the system application. The camera utilized for image capture of the test sequence may be subject to upgrades as technical advances permit higher sensitivity and reduced exposure time (Mentzer, 2011).

Further development of flash X-ray cinematography includes the possibility of using directly illuminated CCD-based detectors, UV scintillating screens, and X-ray diffraction apertures to control and or reduce X-ray exposure regions (Helberg et al., 2006; Cloens, 2006). Further advances in spatial metrology methods (such as point cloud algorithms) will further enhance the analytical capabilities for flash X-ray cinematography.

TERAHERTZ IMAGING

Terahertz imaging (Nova Science Publishers, 2017) is a useful emerging technology developing at Applied Research and Photonics by Dr. Anis Rahman and colleagues (Rahman and Rahman, 2017) for characterizing helmet materials for improved head protection from mTBI. A high-sensitivity, high-speed terahertz dynamic reflectometer is used to measure reflectance kinetics spectra associated with impact events in real-time. Critical parameters related to blunt trauma criterion are computed and other important parameters are extracted from the reflectance kinetics spectrum including dynamic deformation, propagation velocity, final relaxation position, and any delamination characteristics. Kinetics spectra are utilized to compute the deformation profile and the propagation velocity profile via á priori in-lab calibration (Rahman et al., 2014).

Engineering of helmets for improved head protection is an ever-challenging task requiring the characterization of materials to help develop better protection from a variety of impact events. Current characterization methods are limited in arriving at the precise information regarding helmet protection, thereby hindering effective characterization. A high-sensitivity terahertz dynamic reflectometer (TDR) may be used to measure the surface deformation characteristics in real-time (in-situ) and also at post deformation (ex-situ). Real-time measurements can capture layered material deformation kinetics due to impact. Since terahertz radiation can penetrate many composite materials, it produces a clearer picture of the internal material layers of composite laminates and their delamination. A number of crucial parameters can be extracted from the kinetics measurement, such as deformation depth, deformation propagation velocity, and final relaxation position, including vibrational motions of helmets due to impact. In addition, for nonmetallic substrates, a transmitted beam may be used to calibrate mass loss of the laminate layers due to impact. This provides computation of force and energy of impact in real-time.

Current technologies offer limited sensitivity to certain important parameters, such as kinetics and dynamic mass loss that are crucial to fully quantify impact events. Terahertz interaction with materials provides much higher sensitivity because the probing beam is sensitive to vibrations of molecules as a whole as opposed to just a bond or its torsion (Rahman, 2011). Existing methods such as digital image correlation (DIC) (Reu and Miller, 2008; Mentzer 2011) suffer from critical limitations because they are not able to see changes of the internal layers. While X-ray can penetrate, one can only view a frozen-in-time picture after the event has already finished (ex-situ); and dynamic information is not readily available. DIC offers time-evolution of the deforming surface (Rahman and Mentzer, 2012), but is not capable of providing any information regarding delamination or interior change in mass. Moreover, DIC lacks very fine resolution necessary for in-situ inspection of time-dependent changes in materials. Terahertz reflectometry can address these deficiencies.

Unlike other forms of energy that can be used for probing materials and physiological events, such as X-ray, UV, visible, infrared, and ultrasonic, terahertz radiation (T-ray) provides unique advantages in that the T-ray can penetrate many materials and biological tissues, allowing sub-surface interrogation in a nondestructive and noninvasive fashion. T-ray is nonionizing; therefore, it does not cause radiation damage.

Consequently, T-ray-based reflection beam kinetic spectra produce a clearer picture of internal layers of composite laminates and delamination than simple surface measurements. Additionally, a transmitted beam kinetics spectrum may also be used to probe any mass loss of the material due to evaporation at impact or mass gain from projectile fragments. The real-time kinetic spectrum can be used for computation of force and energy of impact that in turn may be exploited for the evaluation of trauma conditions from the Sturdivan criterion (Sturdivan et al., 2005). Currently available technologies offer limited sensitivity to certain important parameters, such as dynamic mass loss, that are crucial to fully quantify a ballistic event.

We used nonrelativistic equations of motion for interpreting the kinetics spectrum for ballistic event characterization of a less-than-lethal impact—the so-called "blunt criterion" (Sturdivan et al., 2005). The experimental setup and the calibration procedure for quantifying deformation from the kinetics spectra, along with calculation of mass loss due to ballistic impact, are described as follows (Rahman and Mentzer, 2012).

Details of terahertz generation are reported in Rahman and Rahman (2012). A dendrimer dipole excitation (DDE) based terahertz source (Rahman 2011) was used where an electro-optic dendrimer with high second-order susceptibility was pumped by a suitable laser system. It is capable of generating continuous wave T-rays with relatively high power (>10 mW). The DDE-based terahertz system is attractive because it eliminates the use of a femtosecond pulsed laser with alternative generation schemes, making it more cost-effective as well as tunable for both bandwidth and output power.

QUANTITATIVE ANALYSIS OF A BALLISTIC EVENT

In the case of a helmet, an important quantity is the available energy for potential impact to the player's head, leading to trauma or injury. Therefore, an important requirement is the quantification of this energy, E_{trauma}. At the point of impact, the kinetic energy, E_k, is simply:

$$E_k = \frac{1}{2} m_p V_p^2, \tag{3.1}$$

where, m_p is the mass of the colliding player, and V_p is the impact velocity. Sturdivan et al. (2005) indicated that the physical quantity properly expressing the capacity to do work on tissue and cause damage from blunt impact is "energy." He expressed the blunt criterion (BC) as a measure to predict head injury from blunt, less-than-lethal impacts, as

$$BC = \ln\left(\frac{E}{T * D}\right), \tag{3.2}$$

where E is the impact kinetic energy in Joules, D is the diameter of the impact contact in centimeters, and T is the thickness of the skull in millimeters. Impacts to the outside of a helmet cause the inside of the helmet to deform (bulge) inwards, thus imparting energy to the player's head. It is this energy that causes trauma or injury; which is less than the impact kinetic energy (E_k) of the impact on the helmet's outer skin. Equation (3.2) therefore takes the form,

$$BC = \ln\left(\frac{E_{trauma}}{T * D}\right). \qquad (3.3)$$

The deformation propagation velocity obtained from the kinetics data gives the velocity profile from which V_{max} for the helmet interior surface is calculated. However,

$$E_{trauma} = \frac{1}{2}m_{eff}V_{max}^2, \qquad (3.4)$$

where, m_{eff} is the effective mass of the deformed portion of the helmet. Knowing m_{eff} one can quantify the energy of BC. Note that neither DIC nor X-ray can determine m_{eff} because, while the density may be approximated from the known material properties and the effective area may be estimated from the post-event device under test, the effective mass of the trauma-generating volume is still not determined. Since many helmets are made of multi-layered material, one needs to know delamination characteristics and possible loss of material during impact. Thus m_{eff} must be determined experimentally. Since terahertz radiation penetrates the helmet material, it is possible to determine any mass loss/gain due to impact. In this case, calibration of material mass as a function of THz transmission must be done á priori.

In light of the foregoing, the total energy delivered by impact then comprises two components: $E_k = E_{trauma} + E_{diss}$, where E_{diss} is the energy dissipated by the helmet material (see Figure 3.3). While it can be easily assumed that $E_{diss} = E_k - E_{trauma}$, the nature of E_{diss} has some interesting connotations. It is hypothesized that helmet impact may generate shock waves (Stoughton, 1997), which may also contribute to trauma. In either case, the net effect of impact, under BC criteria, is the trauma-generating energy E_{trauma}, and thus E_{trauma} is still dependent on m_{eff}, which must be measured. THz reflectometry provides an opportunity to quantify m_{eff}.

Experimental Validation

Figure 3.4 shows the experimental setup for real-time in situ kinetics measurements. An electro-optic dendrimer-based terahertz source generates terahertz radiation up to ~35 THz (Rahman and Rahman, 2012). The source and the detection units remain stationary and are oriented at angle $\Theta = 35°$ or any suitable angle, such that the impact path is clear from entering the instruments. As the target moves from its initial position along the x-axis, both Θ and the deformation (S) become position dependent; $\Theta \rightarrow \Theta(x)$ and $S \rightarrow S(x)$. Thus, the reflected power is a function of x that can be described by Fresnel's law (Piesiewicz et al., 2007). A controlled measurement of power vs. displacement then serves as a measure of deformation for a given surface corresponding to the measured kinetics spectrum at a specific angular orientation and other alignment conditions.

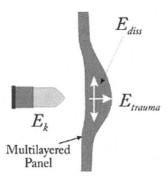

FIGURE 3.3   The incident energy is the sum of the trauma-generated energy and the dissipated energy: $E_k = E_{trauma} + E_{diss}$. (Courtesy of Applied Research and Photonics.)

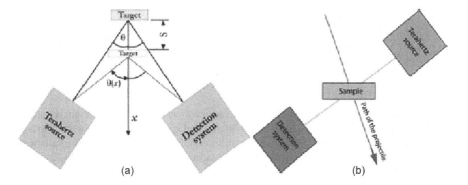

(a)   (b)

FIGURE 3.4   Experimental setup of TDR. (a) Reflection calibration: the terahertz source and the detection unit remain stationary while the target (football helmet) undergoes sudden deformation by absorbing impact. The beam from the source is incident at an angle (35°) on the target and reflected back into the detection system. (b) Experimental setup for mass vs. transmission calibration. (Courtesy of Applied Research and Photonics.)

DEFORMATION CALIBRATION

Deformation calibration involves measuring the reflected energy as a function of known deformation at a given distance and at a known angle of incidence. Figure 3.5 shows the calibration curve for three different target materials. Here the reflected power has been measured as a function of displacement. Each of these curves will serve as a "look-up" table for quantifying the deformation of the corresponding material for a given impact kinetics spectrum (see Figure 3.6). The kinetics is recorded and the deformation is then read off the corresponding calibration curve. When a helmet is placed at a fixed position within the limits of its calibration, the measured power remains unchanged. A sudden displacement of the helmet or a deformation on a localized section where the THz beam is incident causes the power to drop proportional to the displacement and then becomes steady again at the new position. The power vs. time curve (kinetics spectrum) allows quantification of the deformation from this curve.

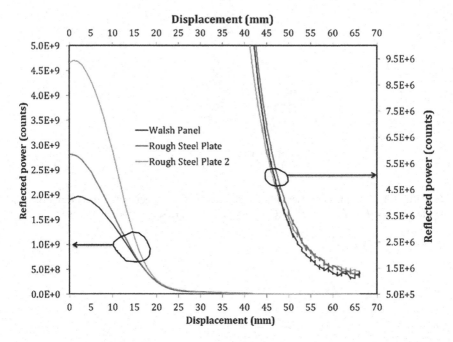

FIGURE 3.5 Calibration of deformation (displacement) vs. power. While the curves on the left Y-axis tend to go to zero at displacement 30 mm and above, however, when the Y-axis is expanded (right Y-axis), the reflected power is still a rapidly varying function of displacement. This indicates that the calibration is valid for displacements up to at least 60 mm. (Courtesy of Applied Research and Photonics.)

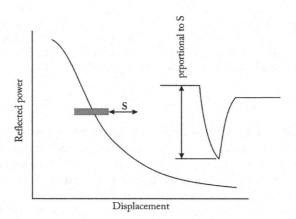

FIGURE 3.6 When a helmet is placed on the displacement curve shown, the measured power remains steady as long as the target remains fixed. A sudden displacement (deformation) of the helmet causes the power drop proportional to the displacement and then becomes steady at the new position. Therefore, the power measured in real-time generates a kinetics curve (inset) from which corresponding displacement can be quantified. (Courtesy of Applied Research and Photonics.)

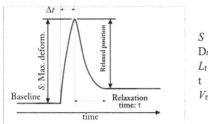

FIGURE 3.7 Definition of parameters extracted from the kinetics spectrum. (Courtesy of Applied Research and Photonics.)

MASS CALIBRATION

Once the deformation is read from the kinetics spectrum, several quantities may be extracted, which can be used to further characterize different candidate materials. These parameters are illustrated in Figure 3.7. The principle used is the Beer–Lambert law, which is normally used to determine the concentration dependence C of a solute in a solvent from absorbance data A: $A = \varepsilon l C$, where l is path length and ε is the extinction coefficient or molar absorptivity. In our case all material parameters may be assumed fixed, with the path length l being replaced by mass m due to delamination. Thus, the transmittance T is proportional to the variation ion path length or, equivalently, the mass change. Therefore a priori measurement of $T(t)$ versus known m can be used to compute the change of mass: $T(t) = \varepsilon m \rho$ where ρ is the density and t is the observation time.

The target, THz source, and detection system are organized so the impact path remains clear. With this orientation the source and detection system are aligned so the detector receives maximum power. A multilayered panel of helmet material is mounted on a fixed platform, placed into the beam path, and the initial power was recorded. Then a cluster of a few layers of the panel was peeled off and transmission was measured again. In this way transmitted power was recorded while other clusters from the panel were successively removed.

For each cluster removed, a small disk was cut out with a diameter approximately equal to the beam spot and its mass measured on a microbalance. Figure 3.8 shows successive layer mass dependence of transmitted power as well as measured power versus measured mass of the disks cut from the peeled clusters. This serves as the calibration for determining mass change during impact. This calibration must be done for a given geometry and material.

As seen in Figure 3.8, transmitted power increases as successive layers are removed from the panel under calibration. The transmitted power is plotted as a function of cumulative mass removed from the panel. The change in mass may be read from the curve or calculated from the fit:

$$Y = 2.11614423 \, X \, 10^7 \, \exp(0.13377147)X \tag{3.5}$$

where Y is the transmitted power and X is mass in grams.

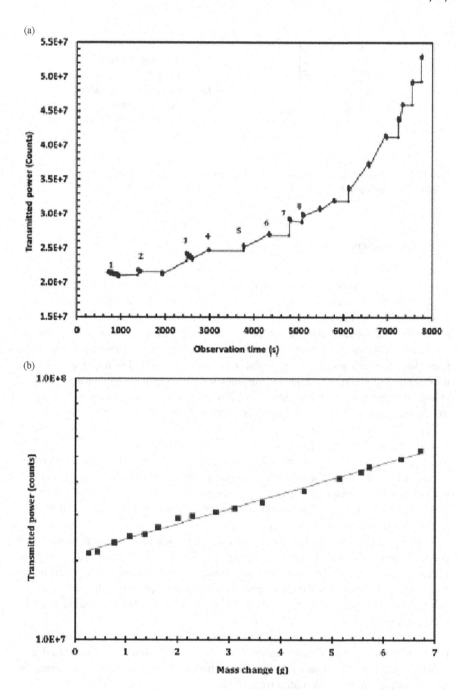

FIGURE 3.8 (a) Transmitted power increases as successive layers are removed from the panel under calibration. (b) The transmitted power is plotted as a function of cumulative mass removed from the panel. The change in mass for this particular panel may be read from this curve or may be calculated from Eq. (3.5). (Courtesy of Applied Research and Photonics.)

COMPUTATION OF PARAMETERS

After the maximum displacement is read-off of the kinetics spectrum utilizing the calibration curve, the deformation propagation profile is calculated from the spectrum. From this profile the next most important parameter to calculate is the speed of deformation. Here the following boundary conditions are utilized. Initially the target is at rest; therefore, the initial velocity is zero. As the deformation propagates, the propagation accelerates and then at the maximum deformation, the velocity is again zero. If the helmet recoils (in the opposite direction), the velocity again increases and then comes to zero when the target stops at the relaxed position. On utilizing Newton's laws for uniformly accelerated motion:

$$S = v_0 t + \frac{1}{2} a t^2 \qquad (3.6)$$

where S is the deformation, v_0 is starting velocity, a is acceleration, and t is time.
Since

$$v_0 = 0 \rightarrow S = \frac{1}{2} a t^2,$$

$$or\, a = 2S/t^2 \qquad (3.7)$$

Knowing a, one can determine v from

$$v^2 - v_0^2 = 2aS \qquad (3.8)$$

Figure 3.9 displays data from a typical experiment and velocity profile. Here the kinetics spectrum was denoised with a curve-fitting routine and then a moving average was adapted to calculate the velocity profile (see Figure 3.10 for definition of parameters). Since the deformation calibration for this sample was not done ahead of time, information available from a simultaneous digital image correlation (DIC) measurement was used to calculate the deformation profile. The velocity profile reveals that the deceleration of the panel after initial acceleration (between time 0.26 and ~0.28 s) exhibits several small accelerations (indicated by the upward shots or kinks in the curve) followed by small decelerations. This is indicative of the clusters of layers moving together as opposed to the whole panel moving; i.e., a few layers forming a cluster and becoming delaminated. Four clusters are visible following the initial deceleration of the panel. When there are distinct slopes before the deformation reaches the maximum, this is indicative of delamination of layers in a cluster within the panel rather than individual layers.

DATA MANAGEMENT AND INTEGRATION

As bioinformatics and ohmic data, along with systems biology approaches to biomarker analysis, proliferate in research labs globally, there is a need to curate, integrate, process, abstract, and learn from the massive entity of big data available to the

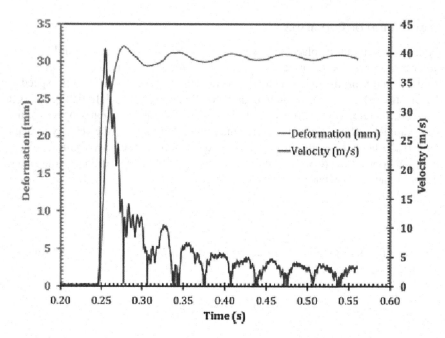

FIGURE 3.9 Typical experiment and velocity profile. Deformation profile (left Y-axis) calculated from the kinetics spectrum with a depth of 34 mm (from DIC under identical conditions). Calculated velocity profile of deformation is shown at right (Y-axis). (Courtesy of Applied Research and Photonics.)

neuroscientist. Textual data must be processed and analyzed along with redundant or uncertain data to create usable knowledge. Machine learning can then provide models for precision medicine, prediction, and association (Kobeissy et al., 2019).

Systems biology has emerged

as an integrative and holistic approach, aiming to assimilate knowledge through relatively large and complex amounts of data generated by high-throughput biological applications and tools … facilitated by the increased availability of novel high-throughput screening techniques such as whole genome sequencing proteomics, and next-generation sequencing (NGS) … all serving to ultimately transform heterogeneous data into useful knowledge that can influence healthcare and biomedical research. (Kobeissy et al., 2019)

Big data promise to deconvolve the pathological processes, severity spectrum, and relationships between mTBI and TBI to those of Alzheimer's and CTE, and to further discovery of biomarkers, personalized therapeutics, image analysis, and legacy data integration (Vo-Dinh, 2013; Vo-Dinh and Askari, 2001). A number of NINDS, DoD, and other organizations, including the Federal Interagency Traumatic Brain Injury Research Informatics System (FITBIR), provide tools for sharing TBI data across organizations, using common data elements and establishment of links to related CNS research (Kobeissy et al., 2019). The common reductionist approach for clinical trials may prove less effective in addressing heterogeneous diseases like

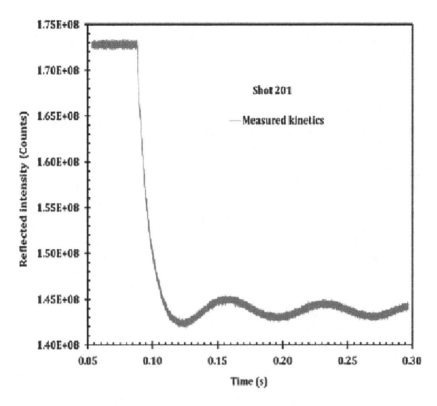

FIGURE 3.10 Another example of kinetics spectral analysis of an impact. Deformation profile (left Y-axis) calculated from the kinetics spectrum with a depth of 34 mm (from DIC under identical conditions). Calculated velocity profile of deformation is shown (right Y-axis). (Courtesy of Applied Research and Photonics.)

mTBI than a systematic integrative analysis that includes genes, proteins, and behaviors (Arrowhead Publishers, 2014). Nonlinear principal components analysis relates observed variables to uncorrelated principal components and provides nonlinear relationships between variables.

COMPARATIVE EFFECTIVENESS RESEARCH (CER)

Comparative Effectiveness Research (CER) is defined as "the generation and synthesis of evidence that compares the benefits and harms of alternative methods to prevent, diagnose, treat, and monitor a clinical condition to improve the delivery of care" (Arrowhead Publishers, 2014). Four unique features of TBI make CER a feasible approach for addressing unknown certainties concerning best clinical practice (Mass et al., 2012):

1. Large between-center differences and between-country differences in outcome and management
2. Robust covariates and validated prognostic models developed in various collaborations to adjust patient characteristics that affect outcome

3. Random-effect models are available to analyze differences between centers
4. Ongoing data management efforts provide standardization of data collection and coding of variables

REFERENCES

Arrowhead Publishers. 2014. *Traumatic brain injury-therapeutic and diagnostic pipeline assessment and commercial prospects.* Chanhassen, MN: Arrowhead Publishers and Conferences.

Cloens, P. 2006. *X-ray imaging instrumentation.* Grenoble, France: ESRF.

Helberg, P., S. Nau, and K. Thoma. 2006. *High-speed flash x-ray cinematography.* ECNDT TH.1.3.2. Fraunhofer Institute for High Speed Dynamics, Ernst-Mach-Institut.

Karsten, M. and P. Helberg. 2005. Computed tomography of high-speed events. *Proc. 22nd International Symposium on Ballistics 2*, Vancouver, BC, Canada.

Kobeissy, F., A. Alawieh, F. Zaraket, K. Wang, eds. 2019. *Leveraging biomedical and healthcare data: semantics, analytics and knowledge.* London: Elsevier.

Mass, A., D.K. Menon, H.F. Lingsma, et al. 2012. Re-orientation of clinical research in traumatic brain injury: report of an international workshop on comparative effectiveness research. *J. Neurotrauma.* 29(1):32–26.

Mattsson, A. 2007. New developments in flash radiography. *Proc. SPIE 27th International Congress on High-Speed Photography and Photonics.* 62790Z.

Medina, C.I., M. Pervizpour, S. Pamukcu, and M.A. Mentzer. 2012. Temperature dependent dynamic properties of oily clay. *GeoCongress GSP 225, ASCE*, 2283–2291. doi:10.1061/9780784412121.234

Mentzer, M.A. 2011. *Applied optics, fundamentals and device applications—nano, MOEMS, and biotechnology.* New York: CRC Press Taylor and Francis Group.

Mentzer, M.A., D.A. Herr, K.J. Brewer, N. Ojason, and H. A. Tarpine. 2010. Detector development for x-ray imaging. *Paper presented at SPIE Photonics West*, January.

National Institute of Standards and Technology Physics Laboratory. 2010. *Photon cross sections, attenuation coefficients, and energy absorptions coefficients from 10 keV to 100 GeV.* NSRDS-NBS 29. U. S. Commerce Department. http://physics.nist.gov/PhysRefData/XrayMassCoef/ElemTab/z06.html.

Nova Science Publishers. 2017. Portions of this section are used and adapted with permission of Nova Science Publishers, Inc., from Rahman, A. and A.K. Rahman. Chapter 8. Terahertz kinetics spectroscopy for laminated helmet materials characterization in *Advances in Materials Science Research.* ISBN 978-1-53612-768-3. Permission conveyed through Copyright Clearance Center, Inc. on March 26, 2020.

Piesiewicz, R.C.J., D. Mittleman, T. Kleine-Ostmann, M. Koch, and T. Kürner. 2007. Scattering analysis for the modeling of THz communication systems. *IEEE Trans Antennas Propagation.* **55**(11):3002.

Prince, J.L. and J.M. Links. 2006. *Medical imaging, signals and systems.* Upper Saddle River, NJ: Pearson Prentice Hall.

Rahman, A. 2011. Dendrimer based terahertz time-domain spectroscopy and applications in molecular characterization. *J Mol Struct.* 1006:59–65.

Rahman, A. and M.A. Mentzer. 2012. Terahertz dynamic scanning reflectometry of soldier personal protective material. *Proceedings Volume 8261, Terahertz Technology and Applications V*, San Francisco, 2012.

Rahman, A. and A.K. Rahman. 2012. Wide range broadband terahertz emission from high c(2) dendrimer. *Proceedings of SPIE – The International Society for Optical Engineering 8261:14-February 2012.* 8261.

Rahman, A. and A.K. Rahman. 2017. Terahertz kinetics spectroscopy for laminated helmet materials characterization. In Wythers, M.C., ed., *Advances in materials science research*. Hauppauge, NY: Nova Science Publishers, Inc.

Rahman, A., A. Rahman, and M. Mentzer. 2014. Deformation kinetics of layered personal protective material under impact via terahertz reflectometry. In Dimensional optical metrology and inspection for practical applications III, edited by K. Harding, T. Yoshizawa, and S. Zhang. *Proceedings of SPIE* 9110.

Reu, P.L. and T.J. Miller. 2008. The application of high-speed digital image correlation. *J Strain Analysis*. 43:673–688.

Shionoya, S. and W.N. Yen, eds. 1999. *Phosphor handbook*. Boca Raton, FL: CRC Press.

Stoughton, R. 1997. Measurements of small-caliber ballistic shock waves in air. *J Acoust Soc Am*. 102(2 Pt. 1):781–787.

Sturdivan, L., D. Viano, and H. Champion. 2005. Analysis of injury criteria to assess chest and abdominal injury risks in blunt and ballistic impacts. *J Trauma Injury, Infect, Crit Care*. 56:651–663.

Vo-Dinh, T. 2013. *Biomedical photonics handbook*. Boca Raton, FL: CRC Press.

Vo-Dinh, T. and M. Askari. 2001. Micro-arrays and biochips: applications and potential in genomics and proteomics. *Curr Genomics*. 2:399.

4 mTBI in the Military and Contact Sports

MILITARY TBI

Recent military conflicts resulted in TBI blast injuries due to improvised explosive devices (IEDs), mortar rounds, rocket-propelled grenades, and suicide bombers. Due to improved body armor, armored combat vehicles, and emergency trauma facilities, more individuals survive events that would result in casualties earlier. Any service member within 50 m of a blast event now receives mandatory systemic evaluation, all of which results in more identifications of mTBIs. Military mTBIs are more likely to accompany PTSD than is the case with sports mTBI, which complicates treatment (Relias Academy, 2020). The neurobehavioral symptom inventory in the clinical practice guidelines for the Veterans Affairs (VA) and Department of Defense (DoD) includes somatic/physical, neurosensory, cognitive, and affective/psychological subtypes (Cifu, 2012).

Deducing the exact nature of blast-induced injury mechanisms and the ensuing pathophysiology presents some unique challenges. Strain wave propagation conditions inside the head have been estimated using viscoelastic theory for isotropic material. Dilatational P-waves exist between 8 kHz and 3 MHz with impact durations less than 1 ms. Distortional S-waves exist between 25 and 300 Hz. Mode conversion at brain boundaries is not important since the P- and S-wave frequency ranges are different. Both P- and S-wave propagation may occur for ballistic impacts as well as for low-velocity high mass impacts (Brands, 2002). High-frequency stress waves and lower-frequency sheer waves propagate through the body, causing injury. When the intensity of the oscillations exceeds the tensile strength of the tissues, the damage threshold has been reached (Kucherov et al., 2012).

Severity of blast TBI (bTBI) depends on the strength of the explosion, reflections of the waves from buildings or vehicles, and distance to the explosion causing the injury (Wang et al., 2002). Peak overpressure is another important parameter related to injury severity, along with duration of the positive wave (Yang et al., 1996). P waves push and pull like sound waves push and pull air. They are also called compressional waves. Particles move in the direction of wave propagation. S waves, also known as secondary waves, move particles up and down or side to side, perpendicular to the direction of propagation. There are also Rayleigh waves that roll across the surface like ocean waves; and there are Love waves—the fastest of the surface waves—that that move particles side to side and produce all horizontal motion.

When a frontal blast wave encounters the head, a range of unique biomechanical and biochemical effects occur, leading to brain injury above certain thresholds. Damage results from the superposition of the waves coupled and transmitted through the tissues. Cavitation produced by negative transient pressures may also contribute to tissue damage.

Progress in mitigating the effects of blast waves from IEDs is hindered by incomplete understanding of blast-induced injury mechanisms, reliance on testing surrogates for injury tolerance and mitigation design, and lack of integration of this understanding into current protective equipment standards (Desmoulin and Dionne, 2009). Enhanced-blast devices use blast as their primary damage mechanism. Such devices include thermobaric, fuel-air, metallized, and reactive surround effects, and alter the blast profile with increased duration and impulse at equivalent peak pressure, thereby transmitting increased energy to the target (Wildegger-Gaissmaier, 2003; DePalma et al., 2005).

This further complicates the design of protective equipment for blast as well as testing to determine injury mechanisms and thresholds. Biofidelic test manikins record only biomechanical parameters with inconclusive links to the injury physiology and to animal models. Head injury criterion (HIC) may be used to predict risk of injury. But they do not account for factors in injury tolerance over time; nor do they account for complex blast parameters such as reflections, reverberation, and interactions of pressure waves within buildings or breached vehicles (Desmoulin and Dionne, 2009). But peak intracranial pressure may represent a reasonable overall predictor of tissue injury (Chen et al., 2009).

Ideally, a model of primary bTBI will consider blast pressure transmitted through brain tissue as well as shock-wave reflection/transmission at tissue interfaces and density discontinuities. It is difficult at this time to identify the most appropriate model for bTBI, as the models should ideally account for both molecular- and cellular-scale effects. More study is needed regarding the cellular and molecular transduction events in primary bTBI to fully understand the causes and mechanisms of bTBI (Chen et al., 2009; Ling et al., 2009). It has been estimated (Ravin et al., 2012; Bowen et al., 1968) that 10 atm peak pressure for a few milliseconds in the skull can result in death [10 atm = 147 psi = 1,010 kPa].

Helmick et al. (2012) listed seven key areas of focus needed to maximize care to service members:

1. Eliminate undetected mild traumatic brain injury through prompt early diagnosis
2. Ensure force readiness and address cultural barriers
3. Improve collaborations with the Department of Veterans Affairs, other federal agencies, and academic and civilian organizations
4. Improve deployment-related assessments
5. Deploy effective treatments
6. Conduct military-relevant and targeted research
7. Enhance information technology systems

TBI AND AMERICAN FOOTBALL

Organized sports and the inherent risk of concussion began in 700 B.C., and clinical symptoms of concussion have been reported since the era of Hippocrates (McCrory and Berkovic, 2001). Awareness of mTBI increased steadily since 2005 when pathologist Bennet Omalu reported extensive damage to a former National Football League (NFL) player. This fueled what has been called an "existential crisis" in American football and a global revolution in concussion safety concerns. A 2015 film entitled *Concussion* starred Will Smith as Dr. Omalu, potrays the pushback by the NFL and controversy over Omalu's findings. NFL doctors attempted to ignore and refute Omalu's conclusions regarding the prevalence and severity of chronic traumatic encephalopathy in football players, sparking additional research and debate. But Omalu certainly receives credit for creating public awareness of the dangers of football-induced mTBI.

Another researcher and major contributor to the understanding of CTE in football players is Ann McKee, a neuropathologist at Boston University CTE Center and chief neuropathologist at the Veterans Administration Boston Healthcare System. McKee amassed an extensive brain bank from several hundred deceased football players. Nearly all the players from the NFL examined had symptoms of CTE. While McKee questions the extent and severity of the problem in football players of all ages, she and Omalu certainly created huge public awareness of symptoms that were recognized since 1928, when a New Jersey pathologist observed the "punch-drunk" syndrome in boxers.

The punch-drunk syndrome became known as dementia pugilistica, and in 1949 neurologist MacDonald Critchley suggested the term "chronic traumatic encephalopathy." An early collaborator of Omalu, neurologist Steven DeKosky, suggests the term "CTE" is a rebranding of the same disease concept and observations of many years. Because CTE could only be diagnosed with certainty after death, subsequent research aimed to improve the definition and classification of CTE. This led to the National Institute of Health further defining the disease signature as clusters of tau protein around blood vessels in the folds of the brain's outer region or cortex.

Researchers hope to further quantify the extent of the risks associated with football traumas as well as with other contact sports such as soccer, ice hockey, and wrestling. It is not clear how the effects of mTBI, leading to CTE, vary among individuals, how best to predict such injuries, and how to prevent or minimize their occurrence. Extensive efforts are underway in the Department of Defense, NFL, NIH, equipment manufacturers, and other organizations either sponsoring or conducting research.

INCREASED AWARENESS OF FOOTBALL-INDUCED mTBI

New rules to lower the incidence of dangerous football events are being instituted each year by the National Football League (NFL) to address growing concerns over the relationship of head traumas experienced routinely in the course of play and leading to the development of CTE. Collisions between players include significant hits to the head, which are acknowledged to lead to CTE. While the ultimate crowd pleasers in football involve touchdowns and spectacular individual play, a huge component of the game is the violence that occurs. Until recently, broken bones, dislocations, tears, and concussions were considered collateral damage as a result of the sport. Furthermore, the players were considered to be fully aware of the dangers inherent in their endeavors. The Monday Night Football program even featured a logo of two helmets crashing together.

Dr. Bennet Omalu, working as a neuropathologist at a Pittsburgh coroner's office, autopsied the brain of a famous football player named Mike Webster. Omalu published his findings in the journal *Neurosurgery*, where he detailed the microscopic observation of excessive tau plaques and neurofibrillary tangles in Webster's brain. The paper was published but received severe pushback from other practicing physicians and scientists, the journal itself, and from the NFL. Omalu was severely criticized in attempts to discredit his work. After all, he was attacking America's greatest sport and challenging the view that hits to the head, concussions, and CTE were unrelated.

Omalu describes his early observations (Omalu, 2017) as follows:

> …many [brain cells] had died and disappeared, and many appeared like ghost cells. A large number of the remaining cells appeared shriveled, as if in the midst of the throes of death. I observed spaces- spongiosis- in the substance of the brain, with shriveled brain skeleton and skeins of brain scars, like a partially demolished building … ugly threads and fibrils of brownish proteins inside and outside the brain cells [tau proteins]. As I examined more slides, I began to notice another type of abnormal protein that should not have been there- the diffuse amyloid protein plaques that were scattered in different regions of his brain and doing the same thing tau proteins did, creating a toxic environment for brain cells.

Following a subsequent autopsy of yet another NFL player's brain, Omalu commented: "…brain sections revealed … abundant amounts of abnormal proteins that are typically found in the brains of eighty-plus-year old individuals who have a certain type of dementia, changes similar to those found in the brains of boxers who suffer from dementia pugilistica or punch drunk syndrome" (Culverhouse, 2012). While tau proteins provide support and nutrient transport in a healthy brain, the abnormal tau formations Omalu observed spread throughout the brain cause cell death and restrict the flow of information along neuronal pathways. This situation did not normally occur in a healthy brain.

This led Omalu to research potential relationships between brain trauma and dementia. While the tau tangles and plaques resembled Alzheimer's disease, the patterns he observed were different. Certainly the "punch drunk" syndrome often seen in boxers suggested that repeated blows to the head result in a form of dementia and lasting pathological condition, and the situation with football was perhaps a less dramatic exhibition of head injury causation. Omalu expected his suggestion of lasting brain pathology caused by football would be well received and considered quite helpful to the sport. He was devastated by the reaction of the medical community and the NFL to his work. His subsequent publications of identical findings in other deceased NFL players heightened the negativity displayed toward him, even as other researchers scrambled to replicate his studies and to furthermore disprove them.

While the NFL conducted their own investigation into Omalu's claims, forming the NFL Disability Committee, they refused initially to acknowledge a link between football and dementia. The committee was staffed with individuals, many of them far less qualified than Omalu in neurology; and they aggressively defended a situation that could and ultimately did become a class action lawsuit. The push back continued for more than a decade as the NFL denied any risk of worsened chronic effects from continued hits to the head. The NFL published more than a dozen papers serving to support their defense (PBS Frontline, 2013).

Football is endemic to America, with a storied tradition dating back more than 140 years. In 1912 it was suggested by some players that perhaps helmets should be worn to play the game. A famous football coach named Glenn "Pop" Warner (1871–1954) advised his college players in Carlisle, PA, that

> playing without helmets gives players more confidence, saves their head from many hard jolts and keeps their ears from becoming torn or sore. I do not encourage their use. I have never seen an accident to the head which was serious, but I have many times seen cases when hard bumps on the head so dazed the player receiving them that he lost his memory for a time and had to be removed from the game. (Omalu, 2008)

Even the US military initially pushed back on the seriousness of mTBI. Colonel Charles Hoge, psychiatrist and director of psychiatry and neuroscience at the Walter Reed Army Institute of Research stated in discussions of his paper (Hoge et al., 2008):

> If you tell a Soldier he's got a mild traumatic brain injury, he'll think, 'Maybe I'm brain damaged'. They don't realize how remarkably resilient the brain is. Then they read in the papers that exposure to a blast leads to brain damage, and that elevates their alarm further.

Hoge contended that the symptoms others associated with mTBI were actually caused by PTSD. He suggested telling soldiers they had concussions rather than calling it mTBI, so they would associate concussion with sports injuries and therefore expect to get better faster.

Such opinions were countered by the media—for example, ABC News anchorman Bob Woodruff's prime time special on TBI and his accounting of problems in VA hospitals treating soldiers with brain injuries—and a RAND Corporation report addressing PTSD and mTBI (RAND Corporation, 2016). The collective conclusion was that Colonel Hoge's downplaying of mTBI was not appropriate (Carroll and Rosner, 2011). Subsequent to this, military funding for mTBI research increased dramatically.

The NFL continued their attack on Omalu. When the league sponsored a conference on CTE in 2007, Omalu was not invited to speak or even to attend. At this point another neuropathologist based in Boston, Dr. Ann McKee, was asked to examine a football player's brain. Informed of Omalu's work, McKee was able to confirm his findings and subsequently became a strong advocate for increased awareness of the issues. She and a colleague held a press conference to discuss their findings during the 2009 Super Bowl football championship in Tampa Bay, Florida, attracting a number of reporters in attendance for the sporting event.

McKee's work led to a meeting request from the NFL to discuss her findings. The NFL was described (PBS Frontline, 2013) as quite defensive, and their in-house experts attacked the suggestion of head trauma leading to CTE. They cited lack of proof of causation or prevalence of the findings, which McKee could not strongly defend, since at that point only six former players' brains had been examined. Meanwhile, a reporter in New York was able to obtain a copy of an internal NFL report on former players. While the report indicated a high prevalence of brain disorders, they claimed that their own study was flawed.

Congress held a hearing later in 2009 in which NFL commissioner Roger Goodell was grilled extensively about what actions the league was taking to address the issue. An analogy was made to Big Tobacco and intense pressure for action by the NFL was exerted. Goodell continually deferred the questions of prevalence and causation to his scientists who were increasingly ill-equipped for denials. The NFL finally relented and openly acknowledged the problem (PBS Frontline, 2013).

The NFL designated McKee's laboratory at Boston University as the "preferred brain bank" and supported the examination of additional NFL player brains. This led to more than 100 similar cases as well as the same pathology in a much younger player with no history of concussion. Finally, a high school player's brain was diagnosed in the same fashion. While McKee was not criticized to the extent that Omalu suffered—she was accused of being more of an activist than scientist, thereby severely overstating the issue—it was clear that Omalu's original findings, postulation of the head trauma—CTE causation, and characterization of the pathology, were correct and well founded. Omalu and McKee presented overwhelming evidence and finally refuted assertions of selection bias, lack of sufficient data, and inconclusive findings.

After endless delaying actions by the NFL a 4,500-player class action lawsuit against the NFL was settled for $765M. The NFL did not admit to guilt and vigorously defended what they knew and when. But public opinion was that they should have known. Continued litigation resulted in the $765M cap on benefits owed to be lifted and increased as more players came to the table to validate their claims.

When Omalu first researched the implications of his observations, he found 37 descriptive terminologies, including CTE, for the pathology. As the overall understanding of the condition continued to improve, he found the term CTE the most appropriate descriptor of his findings and adopted the term for what he observed as the result of head traumas in football. More and more players suffered game- and career-ending blows to the head, and the NFL felt mounting pressure to take a more active role in both acknowledging and diligently acting to protect players and to compensate them for disabilities resulting from the game (Omalu, 2017).

As part of their continued defensive response, the NFL mTBI Committee concluded that concussions on the field result in only minimal brain injury, if at all (Hamberger et al., 2009). A summary of the committee's research was as follows:

> When the immunohistochemical results are extrapolated to professional football players, concussions result in no or minimal brain injury. Repeat impacts at higher velocity or with a heavier mass impactor cause extensive and distant diffuse axonal injury.

> Based on this model, the threshold for diffuse axonal injury is above even the most severe conditions for National Football League concussion.

This essentially states that football is safe—no worries regarding brain injuries. And no problem going right back into the game after a concussion! It's like saying that smoking cigarettes is not harmful! CTE can take 10–20 years to destroy a brain, but the evidence and awareness precipitated by Omalu's diagnosis of Mike Webster's gridiron dementia, and the subsequent label CTE by Omalu, along with the work of others that followed is compelling. In Omalu's original paper in which he discovers CTE in a football player, Omalu (2005) eloquently concludes:

> This case by itself cannot confirm a causal link between professional football and CTE. However, it indicates the need for comprehensive cognitive and autopsy-based research on long-term postneurotraumatic sequelae of professional American football. Empirical, cognitive, and postmortem data on CTE are currently unavailable in the population cohort of professional NFL players. Our report therefore constitutes a forensic epidemiological sentinel case that draws attention to a possibly more prevalent yet unrecognized disease because of the rarity of CNS-targeted autopsies in the cohort of retired NFL players.

Fortunately for the sport and its players, for a participant and those considering participation, and for the benefit of many others, including military patients with head injury, Dr. Omalu has continued his involvement in the field. Along with his colleagues he recently helped identify a chemical radiologic marker called FDDNP used in conjunction with PET scanning to identify the presence of tau and amyloid proteins (Omalu et al., 2018). Omalu also identified a specific gene allele in the player brains he examined, suggesting a possible genetic predisposition to CTE. Dr. McKee also continues to support research in the field as director of the brain bank at BU. She and her colleagues are investigating the cytokine CCL11 in CTE and the prion-like spread of CTE tau.

Biomechanical engineers and equipment manufacturers expend huge efforts to protect players from subconcussions and concussions. But the main role of the helmet remains that of preventing abrasions, cuts, contusions, and lacerations of the face and scalp. No football helmet will protect players from brain injury occurrences. Because children play the game and are not aware of the dangers involved, it is essential that parents and coaches are properly educated. This applies to other contact sports as well, including wrestling, boxing, ice hockey, mixed martial arts, and rugby (Omalu, 2017).

While much of the burden to protect athletes from injury rests with the team doctor, it must be noted that coaches often override decisions of the team doctor. All involved in such decisions should be held liable for their actions. But it is noted that players, like good Soldiers, often beg to be kept in the game. Players in an injury state often lack the cognitive ability to make the judgment for themselves. This dilemma was addressed in a 144-page report from the Congressional Research Service in 2008, which stated "…the NFL and the NFLPA need to make serious efforts … to eliminate the conflict of interest by team doctors who place the financial interests of their teams ahead of players' health."

ANABOLIC STEROID USE BY ATHLETES

An equally daunting challenge to the NFL and athletics in general is the widespread use of anabolic steroids to increase mass and strength. This further complicates the diagnosis and treatment of mTBI. Increased awareness of steroid and human growth hormones by athletes, Congressional hearings, and individual awareness led to the Anabolic Steroids Control Act of 1990. But deceptions and exploitation of athletes continues in this area as well. For instance, only 4% of high school football teams have drug-testing of players (Culverhouse, 2012).

In a 1989 Congressional hearing presided over by Senator Joe Biden, the former NFL players' union executive director Gene Upshaw remarked

> I first must say that it is virtually unanimous why steroid use is pervasive in sports... it has to do with pressure. There is the pressure to earn money; there is the pressure to keep a job; there is the pressure to keep ahead of competition; and there is pressure to win. (Culverhouse, 2012)

Additional challenges for the regulation of sports and protection of the athletes' best interests are emerging. Genetic engineering and manipulation pose a future issue. How will the World Anti-Doping Agency (WADA) address the ethical issues of genetic engineering?

YOUTH FOOTBALL

TBI is the number one cause of death and disability among young adults 19 years old and younger; with a significant number of these due to sports-related head injuries among teenagers (Prins in Ashley and Hovda, 2018). The teenage brain is not fully mature and therefore is more susceptible to a higher rate of trauma than college or professional players. And the skulls of females solidify at age 18—two years after those of males (Culverhouse, 2012). Heading of the ball in soccer often results in traumas from either the ball's impact itself or collision with another player who is also trying to head the ball. It has been suggested that heading be prohibited in all high school games. Progressive neurologic deterioration will start whenever head traumas begin to occur. There is a push for all youth coaches to become certified in first aid, CPR, proper conditioning and skills, and in the dangers of traumatic head injuries. Youth should be educated to respect skills and techniques rather than the violence of play. What price shall we pay for our entertainment (Culverhouse, 2012)?

Unfortunately, despite rule changes designed to make football a safer sport, impact to the head is incidental to the play of the game. In martial arts and boxing, the contact is integral to the play, while in low-contact sports like baseball, basketball, and soccer, head impacts are accidental to the play. It is important to understand and minimize the risks involved in a sport. But the only true means of prevention is not to play the game (Omalu, 2008).

Analogous to limiting the number of pitches thrown by a baseball pitcher, perhaps a measure of cumulative head hits would provide a measure of protection for football players. Guskiewicz at University of North Carolina monitored head impacts with a system called the Head Impact Telemetry System (HITS), which employed sensors inside the helmet and telemetry of force and hit location to a sideline computer. Developed at Virginia Tech, the system demonstrated that college players were routinely exceeding thresholds for injury with 70–75 g forces, while NFL players experienced hundreds of hits in a season at more than 80–120 g (Carroll and Rosner, 2011). Correlated with neuropsychological testing, the system allowed coaches to identify the dangerous technique as part of an effort to reduce the incidence of injury in play.

Community awareness of sports concussions is essential because trainers and doctors are often not available at all games and practices. While only health care professionals can diagnose a concussion, it is important for the general public to recognize and report potential symptoms. To this end, the National Federation of High School sports (https://www.nfhsnetwork.com/company) and the HeadsUp programs serve to educate parents, coaches, and players in sports such as football, ice hockey, lacrosse, and soccer. Further research may identify hidden morbidities not yet captured in current clinical assessments.

RETURN TO PLAY GUIDELINES

A set of return to play guidelines was developed by Robert Cantu at the Boston University School of Medicine (Chin et al., 2011). A concussion grading scale using loss of consciousness (LOC), post-traumatic amnesia (PTA), and post-concussion symptom scale (PCSS) provides a measure of severity of a concussion upon which return to play guidelines are based. An asymptomatic player is free of symptoms for 1 week.

Severity Grading and return to play guidelines are as follows (after Cantu system in Chin et al., 2011).

Grade 1 (mild): No LOC, PTA < 30 minutes, PCSS < 24 hours

Grade 2 (moderate): LOC < 1 minute or PTA ≥ 30 minutes < 24 hours or PCSS ≥ 24 hours < 7 days

Grade 3 (severe): LOC ≥ 1 minute or PTA ≥ 24 hours or PCSS ≥ 7 days

Return to Play Guidelines

Grade 1	First concussion—when asymptomatic for 1 week
	Second concussion—in 2 weeks if asymptomatic
	Third concussion—terminate season, may return next season if asymptomatic
Grade 2	First concussion—when asymptomatic 1 week
	Second concussion—in 1 month if asymptomatic
	Third concussion—terminate season, may return next season if asymptomatic
Grade 3	First concussion—in 1 month if asymptomatic
	Second and third concussion—terminate season, may return next season if asymptomatic.

CONCUSSION COACH

Concussion Coach is an iOS application that can be downloaded at no cost from the National Center for Telehealth and Technology. The platform is used for self-reporting assessments, psychoeducation, and providing instructions for behavioral interventions and resources.

LATEST GUIDELINES FROM THE NCAA

"NCAA member institutions must have a concussion management plan for their student-athletes on file with specific components as described in Bylaw 3.2.4.16 (see Guideline 2i)."

The NCAA Plan (American Association of Neurological Surgeons, 2020)

- Requires that student-athletes receive information about the signs and symptoms of concussions. They also are required to sign a waiver that says they are responsible for reporting injuries to the medical staff.
- Mandates institutions to provide a process for removing a student-athlete that exhibits signs of a concussion. Student-athletes exhibiting signs of concussions must be evaluated by a medical staff member with experience in the evaluation and management of concussions before they can return to play.
- Prohibits a student-athlete with concussion symptoms from returning to play on the day of the activity.
- Requires student-athletes diagnosed with a concussion be cleared by a physician or a physician's designee before they are permitted to return.

The signs of a concussion, according to the NCAA, are as follows:

- Amnesia
- Confusion
- Headache
- Loss of consciousness
- Balance problems
- Double or fuzzy vision
- Sensitivity to light or noise
- Nausea
- Feeling sluggish
- Concentration or memory problems
- Slowed reaction time
- Feeling unusually irritable

LATEST NFL PROTOCOL

Because each player and each concussion are unique, there is no set timeframe for recovery and return to participation under the NFL's current guidelines. The decision to return a player who has a concussion back to practice and games resides with the team physician managing the concussion protocols and is confirmed by an independent neurological consultant (INC), who is consulted specifically for the player's neurological health.

After a player is diagnosed with a concussion, the protocol calls for a minimum of daily monitoring. The player's past concussion exposure, medical history and family history are considered, creating a more complete picture of his health. The protocol progresses through a series of steps, moving to the next step only when all activities in the current step are tolerated without recurrence of symptoms. Communication between the player and the medical staff during the protocol is essential.

The first step is rest. During this time, in addition to avoiding physical exertion, the player is to avoid electronics, social media and even team meetings until he returns to his baseline level of signs and symptoms. The next step introduces light aerobic exercise, which takes place under the direct oversight of the team's medical staff. If aerobics are tolerated, the team physician will reintroduce strength training. The fourth step includes some non-contact football-specific activities, and the fifth step, which is clearance to resume full football activity, comes only after neurocognitive testing remains at baseline and there is no recurrence of signs or symptoms of a concussion.

When the team physician gives the player final clearance, the player has a final examination by the INC assigned to his team. As part of this examination, the INC will review all reports and tests documented through the player's recovery. Once the INC confirms the conclusion of the team physician, the player is considered cleared and is eligible for full participation in the next game or practice.

This protocol allows for players to heal at their own individual rates, includes the expertise of both the team physicians and a neurological consultant and specifically includes an assessment of not only the most recent concussion, but also takes into account the medical history of the player. (American Association of Neurological Surgeons, 2020)

HELMET SAFETY

The GEN II Lightweight Advanced Combat Helmet (LW ACH, 2017; Program Executive Office Soldier Portfolio, 2017) measures linear and rotational accelerations of the helmet as well as blast wave overpressure as indicators of the risk of mTBI. Data is transmitted to a server and analyzed by the Joint Trauma Analysis and Prevention of Injury in Combat (JTAPIC). The intent is to correlate injury events to mTBI.

The Federal Trade Commission investigated companies that manufacture and sell football helmets, including Riddell Sports Group, Inc., Schutt Sports, Inc., and Xenith, LLC for "misleading safety claims and deceptive practices." All three companies agreed to remove potentially deceptive claims about concussion prevention from their advertising. Helmet manufacturers strive to make a better helmet.

Since most hits are off center, the premise is that most head accelerations will be rotational, and that "rotational forces strain nerve cells and axons more than linear forces do" as stated by Robert Cantu of Boston University School of Medicine (Foster, 2012). While helmets do a nice job of preventing skull fractures, it is desired that the helmets would also alleviate injury from rotational forces.

A company named Simbex commercially introduced in 2003 the Head Impact Telemetry System (HITS). The HITS system employs an array of six accelerometers that record location and severity of impacts. Data is transmitted to the sideline for analysis, such as the work at the Virginia Tech Center for Injury Biomechanics. The founding director of the Center Stefan Duma indicates a strong connection between linear and rotational forces (Foster, 2012). Interpretation of the HITS data and validation of the systems sparked much debate about how to use the HITS output for a single event.

Helmet manufacturers continue to experiment with amounts and position of padding, air-filled pockets, and polycarbonate shells. Extra padding to attenuate energy was added in areas identified from NFL films as hit-prevalent areas. The NFL worked with the Canadian firm Biokinetics where a sensing head form was used to test a variety of concepts including protective mouthguards. The Royal Institute of Technology in Stockholm developed the Multidirectional Impact Protection Systems (MIPS) in which a plastic layer fits on the head underneath the padding. When a player takes a hit, the head can float so that the MIPS will eliminate some of the rotational force and prevent a concussion. The technology is made available for licensing to helmet manufacturers (Foster, 2012), despite there being no specific injury threshold for rotational accelerations in agreement with all the stakeholders.

But helmet safety standards such as NOCSAE seem to prevent manufacturers from stepping outside the designs that are approved since that may create legal liabilities. NOCSAE standards were tasked with incorporating rotation so that innovations such as the MIPS could be more readily adapted. Indeed, the helmet and personal protective equipment industry is replete with standards and testing protocols in an attempt to design and test the most relevant parameters for head and body protection (see Figure 4.1).

Experiments have shown (King et al., 2003) that a helmeted head decreased linear acceleration significantly, while angular acceleration was unchanged by the helmet. Since angular acceleration is often considered the main cause of brain injury, the benefit of the helmet for protecting against mTBI is questioned. Shear strain generated from angular acceleration may be the best parameter. In light of this, some conclusions and recommendations were generated in King et al.'s (2003) paper:

- Strain rate and the product of strain and strain rate in the midbrain region appeared to be the best injury predictors for concussion.
- Strain rate is proposed as a cause of brain injury in order to challenge researchers to move away from their focus on either linear or angular acceleration.
- To study injury mechanisms, it is best to focus on brain reaction to complex inputs of linear and angular acceleration.
- The inevitable conclusion is that, if we are able to define tolerance in terms of brain response, we need a computer model to describe this response. Intelligent helmet design will also need such a computer model so that it can afford omni-directional protection to the brain.
- Injury is intimately related to the local response of the brain and not to the global input to the head.

HEAD INJURY PREVENTION TIPS

American Association of Neurological Surgeons (2020) recommendations follow:

Buy and use helmets or protective head gear approved by the American Society for Testing and Materials (ASTM) for specific sports 100% of the time. The ASTM has vigorous standards for testing helmets for many sports; helmets approved by the ASTM

FIGURE 4.1 Personal Protective Equipment testing, verification and certification. (Courtesy of HP White Laboratory, an Intertek Company, with permission.)

bear a sticker stating this. Helmets and head gear come in many sizes and styles, and must properly fit to provide maximum protection against head injuries. In addition to other safety apparel or gear, helmets or head gear should be worn at all times for:

- Baseball and softball (when batting)
- Cycling
- Football
- Hockey

- Horseback riding
- Powered recreational vehicles
- Skateboards/scooters
- Skiing
- Wrestling

Head gear is recommended by many sports-safety experts for:

- Martial arts
- Pole vaulting
- Soccer

Sports Tips

- Supervise younger children at all times, and do not let them use sporting equipment or play sports unsuitable for their age.
- Do not dive in water less than nine feet deep or in above-ground pools.
- Follow all rules at water parks and swimming pools.
- Wear appropriate clothing for the sport.
- Do not wear any clothing that can interfere with vision.
- Do not participate in sports when ill or very tired.
- Obey all traffic signals, and be aware of drivers when cycling or skateboarding.
- Avoid uneven or unpaved surfaces when cycling or skateboarding.
- Perform regular safety checks of sports fields, playgrounds and equipment.
- Discard and replace sporting equipment or protective gear that is damaged.

General Tips

- Wear a seat belt every time, whether driving or riding in a motor vehicle.
- Never drive while under the influence of drugs or alcohol, or ride as a passenger with anybody who is under the influence.
- Keep unloaded firearms in a locked cabinet or safe, and store ammunition in a separate, secure location.
- Remove hazards in the home that may contribute to falls. Secure rugs and loose electrical cords, put away toys, use safety gates and install window guards. Install grab bars and handrails for the frail or elderly.

NFL GAMEDAY OPERATIONS TO IDENTIFY POTENTIAL BRAIN INJURIES

As described to me by an Injury Video Replay Technician for the Cleveland Browns and First Energy Stadium (private conversation, 2019):

In the current NFL operations there are 2 Certified Athletic Trainer (ATC) spotters, which are medical professionals, 2 booth IVRS techs that control tagging potential injuries on the DVR system, and 2 field techs that maintain the sideline boxes that house headsets for the independent (NFL) physicians, a TV for review, and an XBox controller used to change cameras and video playback. When a player goes down, the physicians come over to me and put on headsets so that the 3 IVR personnel for that sideline can communicate with them and give commands on what they want to see, different angles, slow-motion, etc. They then address the player, either in the blue medical tent or in other medical facilities housed within the stadium (X-Ray, team's medical room) or call ahead to the designated hospital/trauma center to inform the assigned physician for that game of the injury, details on how it happened and possible diagnosis so they can be prepared for receiving the player.

According to the (NFL website, 2020)

- Independent certified athletic trainers—ATC spotters—serve as another set of eyes, watching for possible injuries at every NFL game.
- ATC spotters may use a medical timeout to stop the game to remove a player for medical examination.
- Teams aren't charged for timeouts if an ATC spotter stops the game.

NATIONAL RESEARCH ACTION PLAN

Developed by the DoD, Veterans Affairs, and Health and Human Services in response to President Obama's 2012 Executive Order improving access to mental health services for veterans, service members, and military families, the NRAP represents a major milestone in TBI research (Arrowhead Publishers, 2014). It addresses TBI, post-traumatic stress disorder, and other mental health conditions. The aim of the NRAP is "to improve the coordination of agency research into these conditions and reduce the number of affected men and women through better prevention, diagnosis and treatment."

The NRAP vision for accelerating TBI research to improve health care and outcomes was stated as follows (Arrowhead Publishers, 2014).

The inspirational vision for TBI research is to identify evidence-based therapies that are effective in maximizing short- and long-term health and function, community participation and reintegration for persons with TBI in civilian and military populations.

Effective treatments, including rehabilitation treatments, would be personalized to address the specific type of injury and co-occurring conditions (especially substance related), considering patient preferences for care.

A clinically relevant classification system for TBI across the spectrum of injury severities, age, gender, and chronic conditions, including mild single and repetitive injuries would be available to advise patients about their diagnosis, prognosis, and treatment options.

More sensitive, reliable, and efficient tools (gold standards) would be available for evaluating the effectiveness of treatments on an individual's physical, cognitive, and psychosocial functioning and quality of life.

DARPA PREVENT PROGRAM

The goals of the Defense Advanced Research Projects Agency (DARPA) program entitled PREVENT (Preventing Violent Explosive Neurologic Trauma) (Macedonia et al., 2012) were to comprehensively evaluate the physics of the interaction between an IED blast and the brain, and to identify which blast components are associated with neurologic injury and which ones are dominant. Areas of research included molecular and macroscopic causes of bTBI at the cellular, tissue, organ, and system level, pathophysiological injury evolution characterization, and isolation of dominant explosion environment variables and determination of coupled effects on the central nervous system.

PREVENT revealed that mild bTBI can be manifested as an inflammation, which is a natural process of restoration with no tissue destruction. Loads of 25 psi with duration of 6 ms have produced inflammation in live biological specimens. Exposures below 16 psi do not have much bTBI effect on humans (Macedonia et al., 2012).

According to the DARPA website (https://www.darpa.mil/program/our-research/darpa-and-the-brain-initiative)

> The White House announced the BRAIN initiative in April 2013. Today, the initiative is supported by several federal agencies as well as dozens of technology firms, academic institutions, scientists, and other key contributors to the field of neuroscience. DARPA is supporting the BRAIN initiative through a number of programs, continuing a legacy of DARPA investment in neurotechnology that extends back to the 1970s.

REFERENCES

American Association of Neurological Surgeons. 2020. *Website*, accessed March 29, 2020 at https://www.aans.org/Patients/Neurosurgical-Conditions-and-Treatments/Concussion.

Arrowhead Publishers. 2014. *Traumatic brain injury-therapeutic and diagnostic pipeline assessment and commercial prospects.* Chanhassen, MN: Arrowhead Publishers and Conferences.

Ashley, M.J. and D.A. Hovda, eds. 2018. *Traumatic brain injury-rehabilitation, treatment, and case management.* Boca Raton: CRC Press.

Bowen, I.G., E.R. Fletcher, D.R. Richmond, F.G. Hirsch, C.S. White. 1968. Biophysical mechanisms and scaling procedures applicable in assessing responses of thorax energized by air-blast overpressures or by nonpenetrating missiles. *Ann New York Academy Sci.* 152:122 ff.

Brands, D.W.A. 2002. *Predicting brain mechanics during closed head impact-numerical and constitutive aspects.* Eindhoven: Technische Universiteit.

Carroll, L. and D. Rosner. 2011. *The concussion crisis anatomy of a silent epidemic.* New York: Simon and Schuster.

Chen, Y.C., D.H. Smith, and D.F. Meaney. 2009. In-vitro approaches for studying blast-induced traumatic brain injury. *J Neurotrauma.* 26:861–876.

Chin, L.S., G. Toshkezi, and R.C. Cantu. 2011. Traumatic encephalopathy related to sports injury. *US Neurology.* 7(1):33–36, DOI: 10.17925/USN.2011.07.01.33.

Cifu, D.X. 2012. *Managing combat-related mild TBI in the VA polytrauma system of care.* Second Annual Johns Hopkins Traumatic Brain Injury National Conference on Repetitive Head Injury. Baltimore. Johns Hopkins University School of Medicine.

Culverhouse, G. 2012. *Throwaway players.* Lake Forest, CA: Behler Publications, Inc.

DePalma, R.G., D.G. Burris, H.R. Champion, and M.J. Hodgson. 2005. Blast injuries. *N Engl J Med.* 352:1335–1342.

Desmoulin, G.T. and J. Dionne. 2009. Blast-induced neurotrauma: surrogate use, loading mechanisms, and cellular responses. *J Trauma.* 67:1113–1122.

Foster, T. 2012. The helmet that can save football. *Popular Science*, accessed January 18, 2013 at https://www.popsci.com/science/article/2013-08/helmet-wars-and-new-helmet-could-protect-us-all/

Hamberger, A., D.C. Viano, A. Säljö, and H. Bolouri. 2009. Concussion in professional football: morphology of brain injuries in the NFL concussion model-part 16. *Neurosurgery.* 64(6):1174.

Helmick, K., L. Baugh, T. Lattimore, and S. Goldman. 2012. Traumatic brain injury: next steps, research needed, and priority focus areas. *Mil Med.* 177(August supplement):86–92.

Hoge, C.W., D. McGurk, J.L. Thomas, A.L. Cox, C.C. Engel, and C.A. Castro. 2008. Mild traumatic brain injury in U.S. Soldiers returning from Iraq. *New Eng J Med.* 358:453–463.

King, A.I., K.H. Yang, L. Zhang, and W. Hardy. 2003. *Is head injury caused by linear or angular acceleration?* Lisbon: IRCOBI Conference.

Kucherov, Y., G.K. Hubler, and R.G. DePalma. 2012. Blast induced mild traumatic brain injury/concussion: a physical analysis. *J Appl Phys.* 112:104701.

Ling, G., F. Bandak, R. Armonda, G. Grant, and J. Ecklund. 2009. Explosive blast neurotrauma. *J Neurotrauma.* 26:815–825.

LW ACH. 2017. *PEO soldier talks about the advanced Combat Helmet Gen II*, accessed April 4, 2020 at http://soldiersystems.net/2017/03/30/peo-soldier-talks-helmets/

Macedonia, C., M. Zamisch, J. Judy, and G. Ling. 2012. DARPA challenge: developing new technologies for brain and spinal injuries. In Sensing technologies for global health, military medicine, disaster response, and environmental monitoring II; and biometric technology for human identification IX. *Proc. SPIE* 837101-1, Bellingham, WA.

McCrory, P.R. and S.F. Berkovic. 2001. Concussion: the history of clinical and pathophysiological concepts and misconceptions. *Neurology.* 57:2283–2289.

NFL website. ATC spotters, accessed March 3, 2020 at https://operations.nfl.com/the-game/game-day-behind-the-scenes/atc-spotters/

Omalu, B. 2008. *Play hard, die young, football dementia, depression, and death.* Lodi, CA: Neo-Forenxis Books.

Omalu, B. 2017. *Truth doesn't have a side.* Grand Rapids, MI: Harper Collins Zondervan.

Omalu, B., S.T. DeKosky, R.L. Minster, M. Ilyas Kamboh, R.L. Hamilton, and C.H. Wecht. 2005. Chronic traumatic encephalopathy in a national football league player. *Neurosurgery.* 57(1):132.

Omalu, B., G.W. Small, J. Bailes, L.M. Ercoli, D.A. Merrill, K.-P. Wong, S.-C. Huang, N. Satyamurthy, J.L. Hammers, J. Lee, R.P. Fitzsimmons, and J.R. Barrio. 2018. Postmortem autopsy-confirmation of antemortem [F-18] FDDNP-PET scans in a football player with chronic traumatic encephalopathy. *Neurosurgery.* 82(2): 237–246. https://doi.org/10.1093/neuros/nyx536

PBS Frontline. 2013. *League of denial: the NFL's concussion crisis.* PBS Episode 16. October 2013. Accessed March 7, 2020 at https://www.bing.com/videos/search?q=league+of+denial&view=detail&mid=0AF455634041AF2678C10AF455634041AF2678C1&FORM=VIRE

Program Executive Office Soldier Portfolio FY 17. *Generation II helmet sensor (GEN II HS)*, accessed March 16, 2020 at https://www.peosoldier.army.mil/portfolio/Documents/pages.pdf

RAND Corporation. 2016. *Understanding treatment of mild traumatic brain injury in the military health system.* Santa Monica: RAND Corporation. Downloaded April 4, 2020 at file:///C:/Users/markm/AppData/Local/Packages/Microsoft.MicrosoftEdge_8wekyb3d8bbwe/TempState/Downloads/RAND_RR844.pdf

Ravin, R., P.S. Blank, A. Steinkamp, S.M. Rappaport, N. Ravin, L. Bezrukov, H. Guerrero-Cazares, A. Quinones-Hinojosa, S.M. Bezrukov, and J. Zimmerberg. 2012. Shear forces during blast, not abrupt changes in pressure alone, generate calcium activity in human brain cells. *PLoS ONE.* 7:6.

Relias Academy. 2020. Online course *The fundamentals of traumatic brain injury (TBI).* Morrisville, North Carolina: Relias, accessed February 6, 2020 at https://reliasacademy.com/rls/store/?msclkid=974360fd606e157f09973ebfcf5ef713&utm_source=bing&utm_medium=cpc&utm_campaign=Brand%20%7C%20Relias%20Academy%20%7C%202019&utm_term=relias%20academy&utm_content=Relias%20Academy%20%7C%20EM

Wang, Y., Y. Wei, S. Oguntayo, W. Wilkins, P. Arun, M. Valiyaveettil, J. Song, J.B. Long, and
   M.P. Nambiar. 2002. Tightly coupled repetitive blast-induced traumatic brain injury:
   development and characterization in mice. *J Neurotrauma.* 28:2171–2183.
Wildegger-Gaissmaier, A.E. 2003. Aspects of thermobaric weaponry. *ADF Health.* 4:3–6.
Yang, Z., Z. Wan, C. Tang, and Y. Ying. 1996. Biological effects of weak blast waves and safety
   limits for internal organ injury in the human body. *J. Trauma.* 40:S81–S84.

5 Therapeutic Strategies and Future Research

TBI THERAPY

According to Joseph (2011), 1–5% of patients with mild head injury demonstrate focal neurological deficits; 2–3% deteriorate after initially appearing alert and responsive; 3% develop intracranial pressure and hematomas, needing to be removed; and 1% die. Acute management of moderate and severe TBI attempts to minimize secondary injuries with a supportive approach. Guidelines recommend mannitol, barbiturates, nutrition, and prophylactic anticonvulsants, along with neurosurgical interventions to include craniotomy for evacuation of hematomas or edemas (Arrowhead Publishers, 2014). While there are currently no approved neuroprotective agents for improving post-injury recovery, it is possible to support recovery and remove painful symptoms of TBI. While hundreds of clinical trials have been conducted, very few reach the late-stage clinical development phase (Arrowhead Publishers, 2014). Development of diagnostic point-of-care devices and biomarkers will support timely administration of therapeutic protocols.

The cascade of events initiated by primary injury leads to secondary injury such as pathophysiological disruptions, immediate upset of blood and oxygen supply, edema, heightened intracranial pressure, hematoma, anemia, abnormal blood coagulation, hypoxia, infection, and changes in cardiac or pulmonary function. Secondary injury is not immediately evident, but rapid diagnosis of the progression is very important (Arrowhead Publishers, 2014).

SENATE COMMITTEE ON ARMED SERVICES PERSONNEL SUBCOMMITTEE HEARING ON THE STATE OF mTBI CARE AND RESEARCH

At a recent Senate Committee on Armed Services Personnel Subcommittee hearing on the state of mTBI care and research, Mayo Clinic sports neurology and concussion program director David Dodick testified

> Even for those of us who have been examining patients for over 20 years, the signs can be so subtle that they are not picked up on the routine bedside neurological examination… even when the diagnosis of concussion is made, the challenge of managing the [TBI] patient is difficult, because there are no pharmacological agents- not a single one- that has been shown to be effective in improving symptoms or interrupting that secondary injury cascade that occurs. (Basu, 2018)

Senator Thom Tillis called it a "national problem" and that the nation

> must pursue multiple approaches to understand better the chronic effects of mild TBI, including the long-term neurodegenerative problems associated with multiple concussive injuries. From 2000 through the first half of 2017, the Department of

Defense diagnosed over 370,000 servicemembers with TBI. Of that total number of diagnoses, over 305,000 were mild TBIs. We know, however, that mild TBI is not a unique problem within the Department of Defense. (Basu, 2018)

Navy Captain Michael Colston stated that DoD's mandatory screening program

promotes early identification of servicemembers with concussion … remains focused on hard problems around diagnostic clarification, [because of the need] to get return-to-duty determinations, administrative dispositions and medical disability findings right. DoD conducts state-of-the-science research as part of the National Research Action Plan, which coordinates our research priorities with VA and NIH. DoD also collaborates in the national effort to characterize degenerative conditions stemming from subconcussive events or blast exposures. (Basu, 2018)

Christopher Miles of Wake Forest University explained that collaboration

between military and civilian clinicians and researchers in tackling the best way to diagnose and treat concussions is crucial. Although the causes of injuries may be different, though certainly not always, the importance of being able to accurately diagnose and provide the best treatment is the same… [and that development of] an objective test that will help diagnose and guide the management of this condition [is crucial]. Current concussion tools are in use, such as the King-Devick [Mayo Clinic tool: https://kingdevick.com/products/concussion-screening/], but [what is needed is] a gold standard for concussion testing. If an imaging or a blood test similar to what we have for evaluating heart attacks were to be discovered, the evaluation and management tool could be standardized.

Retired Four-Star General and Vice Chief of the Army Peter Chiarelli has been a strong advocate for research funding to find effective treatments and cures for TBI and PTSD. Now an executive with One Mind for Research (https://onemind.org/tracktbi/), Chiarelli feels private funding can propel research faster than government programs. In an interview (Erwin, 2012) he states

It is a shame that the so-called invisible wounds of war get so little attention. Government is too slow. And even if Congress appropriated $100 million, it would end up dividing up the pie into many pieces in order to satisfy multiple constituencies, which would dilute its benefits. [Military equipment suppliers are] one of the groups that I think should help us. I'm not saying they are responsible [for troops being injured] however many of them made a heck of a lot of money … and have a moral obligation to assist these Soldiers and make sure we do everything we can to take care of them. We would like to get some sustained commitment from the defense companies.

It's a win-win for everyone. Rather than each company going down a rat hole in their research, not telling anyone and letting another company go down the same rat hole, companies can work together [and share resources]. The results of the research will not just benefit American soldiers and veterans, but also the much larger population of brain-disease sufferers. The most prevalent wounds of these wars are those we can't see. We don't have good diagnostics and we don't have good treatments.

CLINICAL TRIALS FOR NEUROPROTECTIVE AGENTS

Clinical trials for neuroprotective agents are hindered by the heterogeneity of brain injury etiology, pathology, disease mechanisms, and outcomes. Without mechanistic targeting, many trials aimed to reduce heterogeneity by restricting enrollment criteria, thereby decreasing statistical validity of the trial (Mass et al., 2012). Problems in TBI clinical design include population selection, standardization of treatment and center monitoring, and outcome endpoint identification (Arrowhead Publishers, 2014).

The Neurological Outcome Scale for TBI (NOS-TBI) is one potential tool for outcome assessment (Wilde et al., 2010). The IMPACT mission attempts to optimize trial methodologies and to maximize chances for Phase III trials to demonstrate an effective new therapy. A variety of core measures are suggested for use in assessing global outcomes of a trial (Arrowhead Publishers, 2014).

REHABILITATION

Rehabilitation focuses currently on the alleviation of physical, cognitive, communicative, neurobehavioral, and psychological deficits arising from brain injury. Acquired brain injury is now viewed as a collection of various diseases that may contribute or accelerate a range of neurodegenerative conditions (Ashley and Hovda, 2018). A primary component of brain dysfunction is diffuse axonal injury (DAI), identified as the primary component in 40–50% of all traumatic brain injuries (Meythaler et al., 2001). DAI is specific to brain regions including the parasagittal white matter in the cerebral cortex, the corpus callosum, and the pontine-mesencephalic junction adjacent to the superior cerebellar peduncles (Graham et al., 1987).

Complementary strategies for neuroprotection, neuroactivation, growth promotion, and cell therapies will become routine patient management practice as understanding of the interplay between primary injuries and their consequences to cellular metabolic dynamics, blood-brain barrier integrity, inflammatory responses, and endocrine function is better understood. Interventional and preventive medicine will complement the practice of recovery management, especially with the development of clinical biomarkers for management of improved therapies (Ashley and Hovda, 2018).

During the development of the 2016 version of the VA/DoD Clinical Practice Guidelines (CPGs), the working group identified the following areas for conducting future research:

A. Diagnosis and Assessment
- Long-term outcome studies with a focus on the role of laboratory, imaging, or physiologic testing in the management of and clinical decision-making with a patient more than 7 days following concussion
- Research to improve the diagnostic accuracy of tests for concussion/mTBI in the post-acute period
- Studies that acknowledge the lack of validation of existing case definitions of mTBI and examine diagnostic accuracy of cognitive and neuro-psychological tests for concussion/mTBI
- Examine mechanism-specific physiologic response and associated pathophysiology for which specific treatment and predictive outcome measures may be of value.

B. Treatment
   • Studies of neuroprotective therapies to produce specific treatment rec-
     ommendations or prognostic models based on individual mechanisms
     of injury
   • Studies to address headache management specific to patients with a his-
     tory of mTBI
   • Controlled research to examine vestibular rehabilitation exercises
     in patients with a history of mTBI; define types of dizziness in this
     patient population that will respond positively to specific vestibular
     rehabilitation
   • Head-to-head comparison trials on the difference between tinnitus from
     mTBI versus tinnitus from any other etiology
   • Examine effective treatment interventions for visual dysfunction fol-
     lowing mTBI
C. Care Delivery
   • The role of interdisciplinary/multidisciplinary teams in the manage-
     ment of patients with chronic or persistent symptoms attributed to a
     history of mTBI
   • The efficacy of stepped collaborative care models of treatment deliv-
     ered in primary care settings.

RECOVERY PROGNOSIS

Most patients with mTBI recover within a week to 3 months, while some require up to
a year. Much depends on the nature of the injury, related physical injuries or impairments, and co-morbid conditions such as PTSD (Hou et al., 2011; Silver et al., 2011).
Physical and cognitive rest during the first 30 days of recovery is beneficial. Some early
research indicated taxol may inhibit axon degeneration by stabilizing microtubules
(Smith, 2012).

Apolipoprotein (apo) E is a major genetic risk factor for Alzheimer's disease
(65%–80% of AD patients have one or more apoE4 allele). Perhaps as the allele plays
a role in mTBI, the detrimental effects could be modulated by potential therapies
(Mahley et al., 2012). The apoE receptor ligand reelin mediates signaling in several
molecular pathways and could be an important factor in modulating amyloid and tau
pathologies (Krstic et al., 2012). Reelin receptors apoER2 and VLDLR are part of
normal synaptic plasticity, learning, and memory (Weeber, 2012). Therapies to slow
the rate of decline from tau aggregation for both AD and mTBI are being pursued
but none are currently available. Another potential therapy for mTBI and the secondary injury effects is the gamma-secretase inhibitors. Gamma-secretase blocking can
reduce motor and cognitive deficits and reduce cell loss following TBI (Burns, 2012).

GENETIC DETERMINANTS OF OUTCOME

Sorting the genetic responses to neurotrauma is a complex process. Four broad categories in which genotype could play an important role in outcome after trauma
(McAllister, 2011) follow:

1. Modulation of injury extent
2. Response, repair, and recovery from injury
3. Preinjury traits and cognitive capacity
4. Links to neurobehavioral disorders

Each of these categories involve multiple genes and complex polygenic control. It should be noted that while the individual contribution of an allele could be small, the effects of multiple alleles are quite significant (McAllister, 2011). Possible future therapies could utilize microRNAs (miRNAs), which comprise 20–24 nucleotide molecules that epigenetically regulate cellular function (Bhalala, 2015; Bashor et al., 2008).

MEMS, MOEMS, NANO- AND BIO-NANOTECHNOLOGIES

MEMs and nanotechnology promise useful applications in the neuroscience arena, ranging from point-of-care diagnostics to therapeutic opportunities. The simple distinction between the terms MEMS and nanotechnology is the size of the devices: MEMS devices typically range between millimeters down to microns, and nano devices are on the nanometer scale. Quoting a set of definitions from (Small Tech Consulting, n.d.):

> MEMS is the integration of a number of microcomponents on a single chip which allows the microsystem to both sense and control the environment. The components typically include microelectronic integrated circuits (the "brains"), sensors (the "senses" and "nervous system"), and actuators (the "hands" and "arms"). The components are typically integrated on a single chip using microfabrication technologies similar to those used for integrated circuits. Nanotechnology takes advantage of the observation that at the nanoscale, properties of materials change. Nanotechnology is that array of technologies that use properties of materials that are unique to structures at the nanoscale.

A Richard Feynman presentation in 1959 discussed material fabrication "from the atom up," leading to a debate regarding the concept of molecular manufacturing and molecular nanotechnology (MNT). The classic Drexler–Smalley debate focused on the feasibility of constructing molecular assemblers (Bueno, 2004). Multidisciplinary efforts continue to pace the complementary fields of MEMS, MOEMS (optical MEMS), nano-, and bio-nanotechnology—with exponential growth upon exponential growth, advancing toward Ray Kurzweil's projected "singularity" in the year 2040. Nanomaterials for biology now include quantum rods and dots (CdSe, CdTe, CdHgTe), nonheavy metal-based quantum dots (InP, $CuInS_2$, silicon, magnetic nanoclinics, silica nanoparticles, metallic (gold and silver) particles and rods, and rare earth doped nanophosphors (Er and $Tm:NaYF_4$) (Prasad, 2003).

A significant subsegment of the MEMS market is that of MOEMS—optical MEMS, partly a result of the fusion of computing and signal processing with photonics and communications, including the marriage of optics and semiconductor micromachining and devices, plays heavily in lab on the chip concept for biomedicine. Certainly, the time is ripe for the convergence of MEMS, nanotechnology, and biotechnology. Perhaps MOEMS devices interacting with biological systems will manipulate nanoparticles in applications for drug delivery, gene therapy, and commercial labs on a chip (Mentzer, 2011).

MOEMS advanced rapidly in part due to the telecommunications markets demand for devices such as add-drop multiplexers, optical cross-connect switches, variable optical attenuators, modulators, dense wavelength division multiplexing, wavelength converters, dynamic gain equalizing filters, tunable filters, and advanced packaging technologies. More than 50 start-up ventures in the late 1990s complemented the efforts of the large telecommunication manufacturers in the acceleration of such market applications.

Significant applications are emerging in the related field of biomimetics, involving biologically inspired synthetic materials, which point to the construction of genetic networks from and within cells, along with intelligent implantable sensors. Microfluidic MEMS constructs will provide processing of antibodies, peptides, metabolic markers, and monitors. DNA may even provide the backbone for computer logic chips. Biosensors mimicking mammalian physiological behavior, fabricated from completely synthetic abiotic materials, can be programmed to sense, respond, and adapt to requirements of living systems—along with providing the basis for a range of chemical and biological sensing systems (Valdes et al., 2009).

With the emergence of multidisciplinary fields such as bioinformatics, mathematical biology, computational biological modeling, biophysics, etc. we should continue to see the cataloging and relational systems approach to biology and the sciences as part of the exponential "Kurzweilian" growth characteristic of technological progress. Crystal growth techniques for the III-V GaAs semiconductor material system are analogous in many respects to growth techniques used to nucleate the precipitation of protein crystalline structure from aqueous solution (Mentzer, 2011).

Quoting D.A. LaVan and R. Langer, MIT researchers from the NSF Symposium in 2001:

> While some may dream of nanorobots circulating in the blood, the immediate applications in medicine will occur at the interfaces among … nanotechnology, micro-electronics, microelectromechanical systems (MEMS) and microopticalelectro-mechanical systems (MOEMS)…. The bounty will not be realized until those trained in these new paradigms begin to … address basic medical and scientific questions.

MEMS AND NANOTECHNOLOGY

The following discussion of micro-electro-mechanical systems (MEMS) and nanotechnology is provided courtesy of Dr. Michael Huff of the MEMS and Nanotechnology Exchange (See: http://mems-exchange.org) at the Corporation for National Research Initiatives.

MEMS is the integration of mechanical elements, sensors, actuators, and electronics on a common silicon substrate through microfabrication technology. While the electronics are fabricated using integrated circuit (IC) process sequences (e.g., CMOS, Bipolar, or BICMOS), the micromechanical components are fabricated using compatible "micromachining" processes that selectively etch away parts of the silicon wafer or add new structural layers to form mechanical and electromechanical devices (see Figure 5.1).

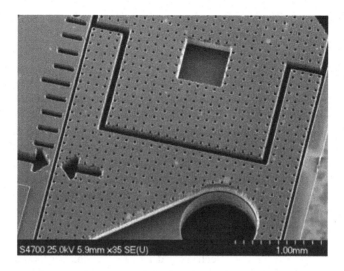

FIGURE 5.1   MEMS structure. (Courtesy of the MEMS and Nanotechnology Exchange.)

MEMS promises to revolutionize nearly every product category by bringing together silicon-based microelectronics with micromachining technology, making possible the realization of complete systems-on-a-chip. MEMS is an enabling technology allowing the development of smart products, augmenting the computational ability of microelectronics with the perception and control capabilities of microsensors and microactuators, and expanding the space of possible designs and applications.

Microelectronic integrated circuits can be thought of as the "brains" of a system and MEMS augments this decision-making capability with "eyes" and "arms," to allow microsystems to sense and control the environment. Sensors gather information from the environment through measuring mechanical, thermal, biological, chemical, optical, and magnetic phenomena. The electronics then process the information derived from the sensors and through some decision-making capability direct the actuators to respond by moving, positioning, regulating, pumping, and filtering, thereby controlling the environment for some desired outcome or purpose. Because MEMS devices are manufactured using batch fabrication techniques similar to those used for integrated circuits, unprecedented levels of functionality, reliability, and sophistication can be placed on a small silicon chip at a relatively low cost.

There are numerous possible applications for MEMS and nanotechnology. As a breakthrough technology, allowing unparalleled synergy between previously unrelated fields such as biology and microelectronics, many new MEMS and nanotechnology applications will emerge, expanding beyond that which is currently identified or known. Following are a few applications of interest to neuroscience.

BIOTECHNOLOGY

MEMS and nanotechnology have enabled new discoveries in science and engineering such as polymerase chain reaction (PCR) microsystems for DNA amplification and identification, micromachined scanning tunneling microscopes (STMs), biochips for the detection of hazardous chemical and biological agents, and microsystems for high-throughput drug screening and selection.

ACCELEROMETERS

MEMS accelerometers are quickly replacing conventional accelerometers for crash air-bag deployment systems in automobiles, and they are also used in some helmet sensors. The conventional approach uses several bulky accelerometers made of discrete components mounted in the front of the car with separate electronics near the air-bag; this approach costs over $50 per automobile. MEMS and nanotechnology have made it possible to integrate the accelerometer and electronics into a single silicon chip between $5 and $10. These MEMS accelerometers are much smaller, more functional, lighter, more reliable, and are produced for a fraction of the cost of the conventional macroscale accelerometer elements.

MEMS and nano devices are extremely small. For example, MEMS and nanotechnology have made possible electrically driven motors smaller than the diameter of a human hair (see Figure 5.2) but MEMS and nanotechnology are not primarily about size. MEMS and nanotechnology are also not about making things out of silicon, even though silicon possesses excellent material properties, which make it an attractive choice for many high-performance mechanical applications; for example, the strength-to-weight ratio for silicon is higher than many other engineering materials, which allows very high-bandwidth mechanical devices to be realized. Instead, the deep insight of MEMS and nano is as a new manufacturing technology, a way of making complex electromechanical systems using batch fabrication techniques similar to those used for integrated circuits, and uniting these electromechanical elements together with electronics.

FIGURE 5.2 MEMS electrically-driven motor. (Courtesy of the MEMS and Nanotechnology Exchange.)

Advantages of MEMS and Nano Manufacturing for mTBI Instrumentation

MEMS and nanotechnology are extremely diverse technologies that could significantly affect every category of commercial and military product. MEMS and nanotechnology are already used for tasks ranging from in-dwelling blood pressure monitoring to active suspension systems for automobiles. The nature of MEMS and nanotechnology and their diverse applications make it potentially a far more pervasive technology than even integrated circuit microchips.

MEMS and nanotechnology blur the distinction between complex mechanical systems and integrated circuit electronics. Historically, sensors and actuators are the most costly and unreliable part of a macroscale sensor-actuator-electronics system. MEMS and nanotechnology allow these complex electromechanical systems to be manufactured using batch fabrication techniques, decreasing the cost, and increasing the reliability of the sensors and actuators to equal those of integrated circuits. Yet, even though the performance of MEMS and nano devices is expected to be superior to macroscale components and systems, the price is predicted to be much lower.

MEMS technology is based on a number of tools and methodologies, which are used to form small structures with dimensions in the micrometer scale (one millionth of a meter). Significant parts of the technology have been adopted from integrated circuit (IC) technology. For instance, almost all devices are built on wafers of silicon, like ICs. The structures are realized in thin films of materials, like ICs. They are patterned using photolithographic methods, like ICs. There are, however, several processes that are not derived from IC technology, and as the technology continues to grow, the gap between those processes and IC technology also grows.

There are three basic building blocks in MEMS technology, which are the ability to deposit thin films of material on a substrate, to apply a patterned mask on top of the films by photolithographic imaging, and to etch the films selectively to the mask. A MEMS process is usually a structured sequence of these operations to form actual devices.

MEMS and nanotechnology are currently used in low- or medium-volume applications. Most companies who wish to explore the potential of MEMS and nanotechnology have very limited options for prototyping or manufacturing devices, and have no capability or expertise in microfabrication technology. Few companies will build their own fabrication facilities because of the high cost. A mechanism providing smaller organizations with responsive and affordable access to MEMS and nano fabrication is essential.

The packaging of MEMS devices and systems needs to improve considerably from its current state. MEMS packaging is more challenging than IC packaging due to the diversity of MEMS devices and the requirement that many of these devices be in contact with their environment. Currently almost all MEMS and nano development efforts must develop a new and specialized package for each new device. Most companies find that packaging is the single most expensive and time-consuming task in their overall product development program. As for the components themselves, numerical modeling and simulation tools for MEMS packaging require further development. Approaches that allow designers to select from a catalog of existing standardized packages for a new MEMS device without compromising performance would be beneficial.

NANOTECHNOLOGY APPLICATIONS

Figure 5.3 from the Foresight Nanotech Institute (2007) represents the technology roadmap for key application areas of nanotechnology. The chart illustrates technology convergence and rapid progression to a host of diverse applications and implementations. Progress in optical bionanotechnology will follow cross-disciplinary endeavors in achieving some of the most significant breakthroughs in science. Applications in biotechnology include bioimaging, biosensors, flow cytometry, photodynamic therapy, tissue engineering, and bionanophotonics. Breakthroughs in biological signaling, genomics, and biosystems engineering will follow as well.

BIONANOTECHNOLOGY

Nanoscale scaffolds may ultimately be utilized for molecular structuring (ala the classic Drexler vs. Smalley debate regarding molecular building blocks and how they might be manipulated) as well as for regulating information transfer. Ron Weiss's page at Princeton (Weiss, 2005) describes building biological circuits. SPICE electrical circuit simulations are now complemented with bioSPICE for modeling genetic circuits (Bio-Spice, 2020). Biological system simulation programs are available as open source code.

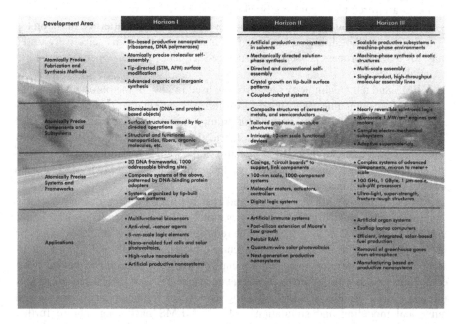

FIGURE 5.3   Nanotechnology roadmap. (Courtesy of the Foresight Nanotech Institute, Palo Alto, CA.)

Molecular recognition is a key objective for a host of multiplexed biosensors technologies in development. These include biomedical MEMS cantilever sensors, surface effects on nanocantilevers, and cancer cantilever sensors. Surface acoustic wave (SAW) sensors prove quite sensitive to adsorbed materials on the device's surface, and a wide range of electrochemical sensors detect free radicals, glucose, and nitric oxide in tissues.

Photonic nanosensors monitor biochemical reactions through conversion of chemical energy to light signals. This is now accomplished for antibody detection inside a single cell (Hornyak et al., 2009). Charge density waves resulting from light impinging on the interface between a thin film and a second medium, called surface plasmons, are utilized to study the formation of self-assembled layers and molecular structures, including DNA and proteins.

BIOLOGY LABS ON A CHIP

Integrated biochips are now commercially available, integrating biotechnology with microelectromechanical systems (MEMS), optics and electronics, imaging, and processing in a miniaturized hybrid system (Vo-Dinh 2003). The term biochip is synonymous in this context with the IUPAC definition of biosensor: "a self-contained integrated device, which is capable of providing specific quantitative or semiquantitative analytical information using a biological recognition element (biochemical receptor), which is retained in direct spatial contact with a transduction element."

The general concept for analysis of a sample (analyte) is for the bioreceptor to provide molecular recognition of the DNA, RNA, or protein molecule, and to effect a signal through a transducer that provides a measurement or characterization output. Biochips are distinguished from microarray assay systems, in that the biochip contains both the microarray and the detector array transduction system. While array plates provide the ability to identify analytes with very high volume, speed and throughput; biochips afford portability, lower cost for small batch processing, and automated analysis algorithms. This is more convenient for applications such as field monitoring or physician's office testing.

Multifunction biochips provide the ability to detect DNA, RNA, protein, and specific characteristics unique to the biological probe composition- such as specific enzymes, antibodies, antigens, and cellular structures (Vo-Dinh and Askari, 2001). Transducers vary widely, such as surface plasmon resonance devices, optical property measurement devices (absorption, luminescence), electrochemical, surface acoustic wave, and other mass measurement devices. Progress in neuroscience and medicine will rely heavily upon the biochip systems currently under development.

APPLICATIONS IN BIOCHEMICAL ANALYSIS, BIOIMAGING, COMPUTING, AND SIGNAL PROCESSING

BIOCHEMICAL ANALYSIS

Jonsson et al. (National Center for Biotechnical Information, 2004) published "A Strategy for Identifying Differences in a Large Series of Metabolomic Samples Analyzed by GC/MS" in which they describe the process for identification of metabolites in a biological system—using standard techniques of organic chemistry including gas chromatography, mass spectrometry, ^{1}H NMR, and data-processing algorithms for rapid analysis of large data sets. Additional organic chemistry analytical methods applied to metabolomics include chromatography, electrophoresis, and mass spectroscopy. Thus, the traditional techniques of organic chemistry are applied to biological systems of metabolites, hormones, and signaling mechanisms in the emergent field of metabolomics.

Metabolomics provides a complete understanding of cell physiology, complemented by the studies of related genomics and proteomics. This approach was first applied to improved understanding of metabolism (Nicholson et al., 1999). It is thought that a systematic cataloging of the human metabolome will provide a baseline for studies of cellular processes and their perturbations.

The Human Metabolome Database (HMDB) database, supported by the University of Alberta and Genome Canada (Human metabolome database version 2.5) contains links to chemical, clinical, and biochemical/molecular biological data, with links to protein and DNA sequences. This systems approach to biological studies promises rapid streamlining of previously tedious processes.

For instance, much additional progress would have resulted in the studies of Dr. Judah Folkman on angiogenesis and its relationship to cancer, had the tools of metabolomics been available a few years ago (Cooke, 2001). Evidence of the growing success of the field is demonstrated in part by growing conferences and attention in the literature to metabolomics—following first mention of the approach about 20 years ago.

Soon perhaps a treatment for cancer epigenetics will be effected through localized solvent physiology changes that reprogram the cancer cell behaviors. If the triggering mechanism for cancer cell programming is identified and reversed, the basis for a cure or reversal of the condition may be possible. Antiangiogenic agents, epigenetic reprogramming—perhaps even structured water with appropriate solvents changing the ion transport of potassium across cell boundaries—may alter sufficiently the folding of proteins during cancer cell processes. It seems that recent breakthroughs (Medical College of Georgia, 2009; University of Illinois at Urbana-Champaign, 2008) in the understanding of DNA methylation may provide the epigenetic key for screening, prevention, and cure of many or all cancers. Analogous approaches in neuroscience promise breakthroughs as well.

BIOIMAGING APPLICATIONS

Optical imaging techniques include fluorescence microscopy, Raman imaging, interference imaging, optical coherence tomography, total internal reflection imaging, multi-photon microscopy, confocal microscopy, and other developing tools, including fluorescence imaging. Fluorescence optical imaging systems include spatial filtering confocal microscopy, spatially resolved localized spectroscopy, polarization and time-resolved fluorescence lifetime imaging (FLIM), and fluorescence resonance energy transfer (FRET). Applications include whole body imaging, drug distribution, protein engineering, and identification of structural changes in cells, organelles, and tissues (Prasad, 2003).

X-ray crystallography provides another means to characterize the atomic structure of crystalline materials. (Rosalind Franklin used the technique to produce the famous demonstration that DNA is helical.) This characterization includes proteins and nucleic acids. While certain proteins are difficult to crystallize, nearly 50,000 proteins, nucleic acids and other biological macromolecules have now been measured with X-ray crystallography (Scapin, 2006; Lundstrom, 2006).

BIOLOGICALLY INSPIRED COMPUTING AND SIGNAL PROCESSING

Interesting switching behaviors occur in proteins (Photonics Media, 2020; Dembski and Wells, 2007). There are aspects of epigenetic programming that may possibly be "switched" with light. If we express such a protein in a cell, we could control aspects of the protein's behavior with light. We must determine what controls and expresses the switching information and what activates it (Haynie, 2001). Some work has been accomplished with cutting and pasting proteins to create light-sensitive kinases. Adding and subtracting amino acid residues modulates the activity (E. Thompson, private conversation 2010; Moglich et al., 2009).

This is analogous to the progression of (electrical) signal processing using combinations of digital logic gates and application-specific integrated circuits; to optical signal processing using operator transforms, optical index, and nonlinear bistable functions through acousto- and electro-optic effects to perform optical computing; to Dennis Bray's "Wetware" biologic cell logic (Bray, 2009), to the bio-philosophical insights of Nick Lane in the last two chapters of his *Life Ascending: the Ten Great Inventions of Evolution* (Lane, 2006; Economist, 2020), where Lane characterizes the evolutionary development of consciousness (Jaynes, 1976) and death (through the mitochondria)—providing the basis for biological logic and how best to modify and control that inherent logic.

We're on the cusp of an epoch—where it is now possible to design light-activated genetic functionality—to actually regulate and program biological functionality. What if, for instance, we identify/isolate the biological "control mechanism" for the switching on/off of limb regeneration in the salamander? Or with the biophotonic emission associated with cancer cell replication—and analogous to destruction of the cavity resonance in a laser, we reverse the biological resonance and switch or epigenetically reprogram the cellular logic? Soon we will engineer time-sequenced reaction pathways to regulate and effect a huge variety of complex biochemical functionalities (Moglich et al., 2009; Lane, 2006; Blum, 1968; Dueber et al., 2009).

The field of optogenetics addresses how specific neuronal cell types contribute to the function of neural circuits—the idea that two-way optogenetic traffic could lead to "human-machine fusions in which the brain truly interacts with the machine rather than only giving or only accepting orders…" (Chorost, 2009). This could be Ray Kurzweil's "singularity" he says is "near" [about the year 2040] in his documentary "Transcendent Man." Device physics and electro-optic/electrical engineering, optical MEMS, and bionanotechnology have converged!

Protein scaffolds bind core kinases that successively activate one another in metabolic pathways engineered to provide biological signal processing functions.

Envision three-dimensional nanostructures fabricated using laser lithography. Then assemble complex protein structures onto these lattices in a preferred manner, such that introduction of the lattice frames might "seed" the correct protein configuration. We then conceive bio-compatible nanostructures for "biologically inspired" computing and signal processing—for biosensors in the military, for example—or for recreating two-way neural processes and repairs.

By spatially recruiting metabolic enzymes, protein scaffolds help regulate synthetic metabolic pathways by balancing proton/electron flux. It seems this represents a synthesis methodology with advantages over more standard chemical constructions. We used to design sequences of digital optical logic gates to remove the processing bottlenecks of conventional computers. But if we look to producing non-Von Neumann processors with biological constructs, we may truly be at the computing crossroad for the "Kurzweil singularity"!

REFERENCES

Arrowhead Publishers. 2014. *Traumatic brain injury-therapeutic and diagnostic pipeline assessment and commercial prospects*. Chanhassen, MN: Arrowhead Publishers and Conferences.

Ashley, M.J. and D.A. Hovda, eds. 2018. *Traumatic brain injury-rehabilitation, treatment, and case management*. Boca Raton: CRC Press.

Bashor, C. 2008. Using engineered scaffold interactions to reshape MAP kinase pathway signaling dynamics. *Science*. 319(5869):1539–1543.

Basu, S. 2018. *DoD, VA still struggle with diagnosing, treating mild traumatic brain injuries*. U.S. Medicine.com website, accessed March 20, 2020 at https://www.usmedicine.com/agencies/department-of-defense-dod/dod-va-still-struggle-with-diagnosing-treating-mild-traumatic-brain-injuries/

Bhalala, O.G. 2015. The emerging impact of microRNAs in neurotrauma pathophysiology and therapy. In Kobeissy, F.H., ed. *Brain neurotrauma-molecular, neuropsychological, and rehabilitation aspects*. Boca Raton: CRC Press.

Bio-SPICE. 2020. *Biological simulation program for intra- and inter-cellular evaluation*, accessed April 5, 2020 at http://biospice.sourceforge.net/.

Blum, H. 1968. *Time's arrow and evolution*. Princeton: Princeton University Press, 159.

Bray, D. 2009. *Wetware: a computer in every living cell*. New Haven: Yale University Press.

Bueno, O. 2004. The Drexler-Smalley debate on nanotechnology: incommensurability at work? *Hyle-Int J Philos Chem*. 10(2):83–98.

Burns, M. P. 2012. γ secretase inhibitor therapy in experimental TBI. *Keystone Symposium on ApoE, Alzheimer's and Lipoprotein Biology and Clinical and Molecular Biology of Acute and Chronic Traumatic Encephalopathies*. Keystone, CO.

Chorost, M. 2009. Powered by photons. *Wired*, November.

Cooke, R. 2001. *Dr. Folkman's war*. New York: Random House.

Dembski, W. and J. Wells. 2007. *The design of life: discovering signs of intelligence in biological systems*. Foundation for Thought and Ethics.

Dueber, J.E., G.C. Wu, G.R. Malmirchegini, et al. 2009. Synthetic protein scaffolds provide modular control over metabolic flux. *Nat Biotechnol*. 27(8):753–759.

Economist. 2008. The origin of life: not that sinister. *The Economist*, April 12, accessed March 23, 2020 at https://www.economist.com/science-and-technology/2008/04/10/not-that-sinister

Erwin, S.I. 2012. Former war commander fighting for funds to combat brain injuries. *National defense*. Arlington, VA: National Defense Industrial Association archives.

Foresight Nanotech Institute. 2007. *Productive nanosystems, a technology roadmap*. San Francisco, CA: UT-Battelle, LLC.

Graham, D.I., J.H. Adams, and T.A. Genneralli. 1987. Pathology of brain damage in head injury. In Cooper, P., ed. *Head injury*, 2nd edition. Baltimore: Williams and Wilkens, pp. 72–88.

Haynie, Donald. 2001. *Biological thermodynamics*. New York: Cambridge University Press.

Hornyak, G., J. Moore, H. Tibbals, and J. Dutta. 2009. *Fundamentals of nanotechnology*. New York: Taylor and Francis Group.

Hou, R., R. Moss-Morris, R. Peveler, K. Mogg, B.P. Bradley, and A. Belli. 2011. When a minor head injury results in enduring symptoms: a prospective investigation of risk factors for postconcussional syndrome after mild traumatic brain injury. *J Neurol Neurosurg Psychiatry*. doi:10.1136/nnp-2011-300767

http://www1.1sbu.ac.uk/water/models.html.

Human metabolome database version 2.5. *Genome Alberta*, accessed April 5, 2020 at http://www.hmdb.ca/.

Jaynes, J. 1976. *The origin of consciousness in the breakdown of the bicameral mind*. Boston: Houghton Mifflin.

Joseph, R. 2011. *Head injury-skull fractures, concussions, contusions, hemorrhage, coma, brain injuries*. Cambridge: University Press.

Krstic, D., A. Maghusudan, P. Vogel, and I. Knuesel. 2012. Modulation of amyloid and tau pathology by the apoE receptor ligand, reelin. *Keystone Symposium on ApoE, Alzheimer's and Lipoprotein Biology and Clinical and Molecular Biology of Acute and Chronic Traumatic Encephalopathies*. Keystone, CO.

Lane, N. 2006. *Power, sex, suicide: mitochondria and the meaning of life*. Oxford: Oxford University Press.

Lundstrom, K. 2006. Structural genomics for membrane proteins. *Cell Mol Life Sci*. 63(22):2597. doi:10.1007/s00018-006-6252-y.

Mahley, R.W., K.H. Weisgraber, and Y. Huang. 2012. Neurobiology of apolipoprotein E in Alzheimer's disease and traumatic brain injury. *Keystone Symposium on ApoE, Alzheimer's and Lipoprotein Biology and Clinical and Molecular Biology of Acute and Chronic Traumatic Encephalopathies*. Keystone, CO.

Mass A., D.K. Menon, H.F. Lingsma, J.A. Pineda, M. Elizabeth Sandel, and G.T. Manley. 2012. Re-orientation of clinical research in traumatic brain injury: report of an international workshop on comparative effectiveness research. *J Neurotrauma*. 29(1):32–26.

McAllister, T.W. 2011. Genetic factors. In Silver, J.M., T.W. McAllister, and S.C. Yudofsky, eds. *Textbook of traumatic brain injury*, 2nd edition. Arlington, VA: American Psychiatric Publishing, Inc.

Medical College of Georgia. 2009. Cancer's distinctive pattern of gene expression could aid early screening and prevention. *Science Daily*. http://www.sciencedaily.com/releases/2009/07/090727110641.htm.

Mentzer, M.A. 2011. *Applied optics-fundamentals and device applications-nano, MOEMS, and Biotechnology*. Boca Raton: CRC Press Taylor and Francis Group.

MEMS and Nanotechnology Exchange. http://www.mems-exchange.org/.

Meythaler, J.M., J.D. Peduzzi, E. Eleftheriou, and T.A. Novack. 2001. Current concepts: diffuse axonal injury-associated traumatic brain injury. *Arch Phys Med Rehabil*. 82:1461–71.

Moglich, A., R. A. Ayers, and K. Moffat. 2009. Design and signaling mechanism of light-related histidine kinases. *J Mol Biol*. 385(5):1433–1444.

National Center for Biotechnical Information. 2004. *A strategy for identifying differences in large series of metabolomic samples analyzed by GC/MS*. U.S. National Library of Medicine. http://www.ncbi.nlm.nih.gov/pubmed/15018577.

Nicholson, J. K., J. C. Lindon, and E. Holmes. 1999. Metabonomics: understanding the metabolic responses of living systems to pathophysiological stimuli via multivariate statistical analysis of biological NMR spectroscopic data. *Xenobiotica.* 11:1181–1189.

Photonics Media. 2020. *Light reveals neuron function.* http://www.photonics.com/Content/ReadArticle.aspx?ArticleID=39843.

Prasad, P.N. 2003. *Introduction to biophotonics.* Hoboken: John Wiley and Sons.

Scapin, G. 2006. Structural biology and drug discovery. *Curr Pharm Des.* 12(17):2087. doi:10.2174/138161206777585201.PMID 16796557.

Silver, J.M., T.W. McAllister, and S.C. Yudofsky. 2011. Textbook of traumatic brain injury, 2nd edition. Arlington, VA: American Psychiatric Publishing, Inc.

Small Tech Consulting. n.d. MEMS and nanotechnology. http://www.smalltechconsulting.com/What_are_MEMS_Nanotech.shtml.

Smith, D.H. 2012. How even a single TBI weaves a tangled web of progressive neurodegeneration. *Second Annual Johns Hopkins Traumatic Brain Injury National Conference on Repetitive Head Injury. Baltimore.* Johns Hopkins University School of Medicine.

University of Illinois at Urbana-Champaign. 2008. Water is 'designer fluid' that helps proteins change shape. *Science Daily.* http://www.sciencedaily.com/releases/2008/08/080806113314.htm.

Valdes, J., E. Valdes, and D. Hoffman. 2009. Towards complex abiotic systems for chemical and biological sensing. *Final report to Aberdeen Proving Ground, ECBC-TR-720.* Approved for public release and unlimited distribution.

Vo-Dinh, T. 2003. *Biomedical photonics handbook.* Boca Raton: CRC Press.

Vo-Dinh, T. and M. Askari. 2001. Micro-arrays and biochips: applications and potential in genomics and proteomics. *Curr. Genomics* 2:399.

Weeber, E.J. 2012. Cognition enhancement by apoE receptor activators. *Keystone Symposium on ApoE, Alzheimer's and Lipoprotein Biology and Clinical and Molecular Biology of Acute and Chronic Traumatic Encephalopathies.* Keystone, CO.

Weiss, R. 2005. Princeton University Department of Electrical Engineering. http://www.ee.princeton.edu/people/Weiss.php.

Wilde, E.A., S.R. McCauley, T.M. Kelly, et al. 2010. The neurological outcome scale for traumatic brain injury (NOS-TBI): I construct validity. *J Neurotrauma.* 27(6):983–989.

Appendix 1
US Patent 9080984 Blast, Ballistic, and Blunt Trauma Sensor

(12) **United States Patent**   (IO) **Patent No.:**   **US 9080984 B2**
Mentzer   (45) **Date of Patent:**   **July 14, 2015**

(54) **BLAST, BALLISTIC, AND BLUNT TRAUMA SENSOR**

(71) Applicant: **U.S. Army Research Laboratory ATTN: RDRL-LOC-1,** Adelph, MD (US)

(72) Inventor: **Mark Andrew Mentzer,** Churchville, MD (USA)

(73) Assignee: The United States of America as **represented by the Secretary of the Army,** Washington, DC (USA)

(*) Notice: Subject to any disclaimer, the term of this patent is extended or adjusted under 35 U.S.C. 154(b) by 0 days.

(21) Appl. No.: **13/737985**

(22) Filed: January 10, 2013

(65)   **Prior Publication Data**

     US 2013/0189795 Al   July 25, 2013

Related U.S. Application Data

(60) Provisional application No. 61/589,005, filed on January 20, 2012.

(51) **Int. Cl.**

GOJN33/53	(2006.01)
GOJN21/78	(2006.01)
GOJN33/543	(2006.01)
GOJN33/58	(2006.01)
GOJN33/68	(2006.01)

(52) **U.S. Cl.**
    CPC   ***GOIN 21178*** (2013.01); ***GOIN 33/5432***
                    (2013.01); *GOIN 33/54366* (2013.01); *GOIN*
                    33/582 (2013.01); ***GOIN 33/6896*** (2013.01);
                    GOIN 2500/00 (2013.01); *GOIN 2800/28*
                    (2013.01); *GOIN 2800/40* (2013.01)

(58) *Field of Classification Search*
    None
    See application file for complete search history.

(56)   **References Cited**

                    U.S. PATENT DOCUMENTS

    6,387,614 Bl*   5/2002 Cheng et al. 435/4
    7,433,727 B2   10/2008 Ward et al.
    2010/0028902 Al*   2/2010 Brown et al. 435/7.1
    2010/0151553 Al*   6/2010 Bjork et al. 435/173.7

                    OTHER PUBLICATIONS

Zhang et al. (*Analytical Biochem.* 1995 vol. 229, p. 291–298).*
Mentzer et al. (*J. Biol. Chem.* 2001 vol. 276, p. 15575–15580).*
Chen et al. (*J. Nanoparticle Res.* 2006 vol. 8, p. 1033–1038).*
Sklar et al. (*J. Biol. Chem.* 1981 vol. 256, p. 4286–4292).*
Ladokhin et al. (*J. Membrane Biol.* 2010 vol. 236, p. 247–253).*
Epand et al. (*Biopolymers* 2003 vol. 71, p. 2–16).*

    * cited by examiner

Primary Examiner—Jacob Cheu
(74) Attorney, Agent, or Firm—Robert Thompson

(57)   **ABSTRACT**

A molecular biosensor is provided including a lipid vesicle and a housing wherein the vesicle is contained in or within the housing and where the housing has a portion capable of transmitting a force generated, external to the housing to the vesicle. The biosensor is used in the process of detecting the presence or absence of an event force such as a blast or blunt force sufficient to produce a medical complication such as a traumatic brain injury.

6 Claims, No Drawings

BLAST, BALLISTIC, AND BLUNT TRAUMA SENSOR

CROSS REFERENCE TO RELATED APPLICATIONS

This application depends on and claims priority to U.S. Provisional Application No. 61/589,005 filed January 20, 2012, the entire contents of which are incorporated herein by reference.

GOVERNMENT INTEREST

The invention described herein may be manufactured, used, and licensed by or for the U.S. Government.

FIELD OF USE

The invention relates to detection of force. More specifically, the invention related to detection of blast or blunt forces such as those impacting a person or vehicle.

BACKGROUND

The field of clinical neurology remains frustrated by the recognition that secondary injury to central nervous system tissue associated with physiologic response to an initial insult resulting from direct blunt force or the percussive forces found in close proximity to a blast source could be lessened if only the initial insult could be rapidly diagnosed or characterized. While the diagnosis of severe forms of such insults damage is straightforward through clinical response testing and computed tomography (CT) and magnetic resonance imaging (MRI) testing, these diagnostics have their limitations in that medical imaging is both costly and time-consuming while clinical response testing of incapacitated individuals is of limited value and often precludes a nuanced diagnosis. In many instances, the instrumentation necessary for these diagnostic procedures is not available in many situations such as in the field. Additionally, owing to the limitations of existing diagnostic tests and procedures, situations exist under which a subject experiences a stress to their neurological condition such that the subject often is unaware that damage has occurred or does not seek treatment as the subtle symptoms often quickly resolve. The lack of treatment of mild to moderate challenges to the neurologic condition of a subject can have a cumulative effect or subsequently result in a severe brain damage event having a poor clinical prognosis.

An analysis of the mechanisms and development of biomarkers related to blast injury is complicated by a deficiency in the number of quality experimental studies, and by the lack of sensitivity and specificity of biomarker-based injury prediction. By the time a biomarker analysis is performed, the subject may be already in a severe and irreversible state. Thus, there is a need for a detection system that can identify the presence or absence of an event severe enough to warrant monitoring or treatment and optionally quantify the extent of trauma an individual has received that will allow for rapid treatment decision-making in the field or in a clinical setting.

Summary

The following summary is provided to facilitate an understanding of some of the innovative features unique to the embodiments of the present invention and is not intended to be a full description. A full appreciation of the various aspects of the invention can be gained by viewing the entire specification, claims, drawings, and abstract as a whole.

A molecular biosensor is provided that allows tor laboratory or field detection of an event, for example an event created by a blunt force or a blast force. Such events are commonly found to be the cause of traumatic injuries, such as traumatic brain injuries. As used herein, an "event force" is any force type suitable to produce or model a traumatic brain injury of any form. Such forces include, but are not limited to, blunt force, ballistic force, shock wave forces illustratively those associated with blast trauma. With traumatic brain injuries and particularly with mild traumatic brain injuries there may be no external signs of injury, which potentially could delay treatment or give an indication that no treatment is necessary, leading to severe, often cumulative consequences. A molecular biosensor and methods provided serve as bio-relevant sensors of traumatic events.

A molecular biosensor includes a lipid vesicle on or within a housing that will not appreciably alter the event force transmitted to a vesicle. A lipid vesicle is optionally tailored to include one or more lipids and optionally other molecules including proteins and cholesterol, among others, to serve as a model similar to the plasma membrane of brain tissue.

Illustrative biochemical components of a lipid vesicle in a biosensor comprise phosphatidylcholine, phosphatidylserine, phosphatidylethanolamine, phosphatidylinositol, sphingomyelin, cholesterol, ceramide, and one or more proteins—for example, integral membrane proteins, NMDA cell surface receptors, rhodopsin, ion channel transporters, and proteins that function to regulate ion transport across the membrane. Biophysical experiments to elucidate the fundamental biochemistry of mild traumatic brain injury or Alzheimer's disease will employ such membrane–protein combinations to study the effects of perturbations of the structural integrity of the sensor constructs. While such model membrane systems have not been constructed for these purposes, a variety of biochemical protocols can be applied in the synthesis of both phospholipid-based liposome assembly, as well as synthetic lipid constructs from which liposome can be assembled, or combinations thereof. The vesicles range in diameter from about 20 to 80 nanometers according to size exclusion chromatography.

In embodiments, the molecular biosensors are useful for the detection, diagnosis, or study of a traumatic event such as a blunt force, blast force, or other force of sufficient magnitude to produce a traumatic brain injury in an animal subject, optionally a human subject. In embodiments, by affixing a molecular biosensor to a subject, an article of protective clothing or another location on a subject, the magnitude of an event can be readily ascertained, which could be used to direct the wearer to medical attention if necessary or for the study of the ability of protective articles to protect a subject.

The biosensors and methods provided address the need for a biologically relevant correlation to traumatic injuries that can be used in either a field or laboratory setting.

DETAILED DESCRIPTION

The following description is exemplary in nature and is in no way intended to limit the scope of the invention, its application, or uses, which may, of course, vary. The invention is described with relation to the nonlimiting definitions and terminology included herein. These definitions and terminology are not designed to function as a limitation on the scope or practice of the invention but are presented for illustrative and descriptive purposes only. While the processes are described as individual steps or using specific materials, it is appreciated that described steps or materials may be interchangeable such that the description of the invention includes multiple parts or steps arranged in many ways as is readily appreciated by one of skill in the art.

A significant technology gap exists in the testing of personal protective equipment for subject individuals and animals, relating to body armor as well as helmet systems, and other protective equipment. Sensors are required to determine the correlation between threats (insults) to the subject so as to provide a means by which protective equipment can be assessed for its ability to protect a subject from a variety of insults and injury, and to optimize the design trade-offs between armor weight, thickness, energy dissipation, stopping power, and the like. This need extends to the widely publicized concerns regarding protection of subjects in conflict or competitive areas, and to the protective measures needed for contact sports, for example, American football.

Recent understanding of the medical conditions, known as traumatic brain injury (TBI), chronic traumatic encephalopathy (CTE), certain aspects of post-traumatic stress disorders (PTSD), and their associated symptoms, further illuminate the need for an improved understanding of the effects of severe trauma to the head, limbs, or torso. Improved prophylaxis includes armor designed to better shield from the insult scenarios—as well as improved post-exposure treatment to alleviate or minimize the short- and longer-term effects of the insults. A gamut of intracranial pathologies results in symptoms including loss of memory, disorientation, angiogenesis, and long-term cognitive disorders.

Clearly a means is required by which the range of insults, including blunt trauma, ballistic impact (often collectively referred, to as B&B), and shock trauma, can be measured with a metric that directly indicates the injury to the body due to an insult thereby directly correlating insult to injury. While a host of sensors have been employed to this end-including pressure sensors, accelerometers, strain sensors, and optical surface measurement methodologies, to characterize the energy impacting the protective armor, and the dissipation of that energy through human tissues and a range of torso and head form anthropomorphic test modules (ATMs) incorporating these sensors, the point and 2D energy characterizations, along with time-resolved networked sensor determinations—have all provided a less than satisfactory correlation of insult to injury. While a host of candidate sensors continue to emerge in the literature (hydro gels, functionalized nanoparticles, photonic crystals, etc.), only the novel sensor concept disclosed herein directly represents the response of human tissues to traumatic insult. Nanotechnology research is replete with examples of self-assembled chemicals forming well controlled supramolecular films and structures, including manipulation of material properties at the atomic level of detail.

Problems with current sensors used in test labs include the lack of repeatable measurement, poor to no correlation, lack of a calibrated response to the range of insults to include ballistic threats; and concurrently, lack of a correlation to any or all of the range of tissue susceptibilities and widely varying vulnerabilities. Test artifacts abound due to a wide range of variables, including threat mass, velocity, total yaw at impact, yaw cycle precession, obliquity at impact, backing material variability, along with backing material inconsistencies, tissue simulant variation, and lack of controlled test protocols proving the repeatability of test metrics. This results in highly conservative limits for penetration depth at prescribed impact kinetic energies, providing only a partially correlated determination of armor suitability and little trade space for the armor designer to effect improvements.

Bullets and fragments cause tissue injury in a number of ways, even if the impact is nonpenetrating. The amount of kinetic energy transferred to the tissues correlates to the severity of the tissue damage, which is determined by four key factors (Cooper, G J., and J. M. Ryan. *Br J Surg.*, 1990; 77:606–610). These include kinetic energy ($\frac{1}{2} mv^2$) at impact, total yaw at impact, shape of the insult, and the characteristics of the target tissue (density, strength, and elasticity). The complete disclosures of the above references are incorporated herein by reference. Nonpenetrating events causing tissue damage mechanisms may be collectively addressed as the disruption of the phospholipid bilayer surrounding the cellular structure of human tissues. Relative damage to tissues correlates directly to tissue densities; such that a measure of lipid bilayer disruption by the threat provides a very direct and novel approach to the lingering problem of insult to injury correlation.

The drawbacks of prior sensor systems and processes of their use are addressed by a physiologically relevant sensor such as those provided. A sensor is provided as well as processes for using a sensor for the detection of a traumatic event and, more specifically, a biosensor for detecting percussive, blunt force, or other trauma. Such sensors can be used to test the ability of protective equipment to protect an individual's brain or other organs from certain traumatic events. Thus, a device is provided that can be used to provide an indication of whether a traumatic event may have caused a traumatic brain injury and to a sensor that can be used to determine the ability of protective equipment to protect against particular threats. The invention has utility as a detector and method of detecting the presence or absence of an event sufficient to produce a traumatic brain injury such as mild traumatic brain injury (mTBI) or other traumatic brain injury.

In embodiments the present invention involves the use of self-assembled liposomes. As used herein, self-assembly of liposomes refers to the thermodynamically stable assembly, in solution, of lipids into the characteristic spherical structures known as liposomes. Hydrophilic acyl tails extend toward one another inside the phospholipid bilayer, and hydrophilic head groups orient toward the aqueous environment, both inside and outside the bilayer.

In embodiments, liposome structures are utilized as a sensor for an event, illustratively an event sufficient to cause mild traumatic brain injury or traumatic brain injury. The sensor is configured and packaged in a manner where the sensors can be affixed to a subject's helmet, body armor, or other personal protective equipment and provide a direct indication of the trauma received at the point of attachment. Packaging configurations will depend on the applications. For perfusion of mimetic brain tissue or gels, the sensor will remain in solution and be perfused into and through brain tissue and surrogates. For example, while using a soldier's helmet or body armor attachment, the liposomes will likely be affixed to gels and the sensing mechanism will occur as the colorimetric changes occur in and on the gels. As an alternative to the gel encapsulation, the liposome solutions will be contained in a pill capsule. In still further embodiments, they can also be contained in quartz cuvettes used to insert the solution directly in the circular dichroism meter. This trauma indicator relates to and is equivalent to the blunt and ballistic trauma, as well as the convolved effect of shock waves associated with blast trauma, received by the human body tissues during equivalent events. The sensor, therefore, represents the first real and direct measure by which insult is correlated to injury. The disruption of the phospholipid bilayer occurring to a human or other subject representing damage to a subject's tissue is the very same disruption measured by a sensor in embodiments of the instant disclosure. Thus, in certain embodiments, a sensor of the present invention is believed to provide a more accurate measure of damage to tissue caused by a blast.

A further beneficial embodiment is to embed the vesicles in a gel and use either confocal or two-photon microscopy (both available in our laboratories) to image the leakage of embedded dyes. Additionally, changes in trans-membrane protein function due to trauma can be implemented in Hodgkin–Huxley-like model neurons by altering parameters representing the fraction, conductance, and reversal potentials of simulated voltage gated channels. Changes in lipid bilayer permeability can be modeled through changes in leakage current conductance.

In embodiments, a molecular biosensor is provided that includes a lipid vesicle (liposome) and a housing. The lipid vesicle is associated with the housing such as being contained within the housing or otherwise attached to or retained by the housing. An event such as a shock wave or blunt force when contacting the housing is transmitted to the lipid vesicle causing alteration in a molecular characteristic of the vesicle that correlates with and indicates the degree, type, duration, severity, or other characteristic of the event force.

A lipid vesicle is sufficiently related to the plasma membrane of a brain cell. Sufficiently related indicates that the lipid vesicle possesses similar lipid and, optionally, protein content to the plasma membrane of a brain cell such that molecular alterations in the vesicle correlate to the damage of a neuron when exposed to an event force. Vesicles that are sufficiently related possess a lipid composition having phosphatidylcholine (PC) and/or phosphatidylethanolamine (PE) as the major lipid component. As such, embodiments of a molecular biosensor include lipid vesicles comprising phosphatidylcholine, phosphatidylethanolamine, or have a lipid content that is 50% or greater, a combination of phosphatidylcholine and phosphatidylethanolamine. Illustrative examples of the lipid content of the plasma membrane are illustratively found in Scandroglio, et al., *J Neurochem*, 2008; 107 (2): 329–338, the complete disclosure of which is incorporated herein by reference. In some embodiments, the amount of lipid (e.g. phospholipid, sphingomyelin, and cholesterol) is 40% represented by the amounts presented by Scandroglio et al., for example about 20–70% molar ratios.

A lipid vesicle is optionally formed from phosphatidylcholine, phosphatidyletha-nolamine, phosphatidylserine (PS), phosphatidylinositol (PI), sphingomyelin, choles-terol, ceramide, or combination thereof. In embodiments, a lipid vesicle comprises about 50–100% phosphatidylcholine or a combination of phosphatidylcholine and phosphatidylethanolamine. Lipid vesicles optionally contain only phosphatidylcho-line as a lipid component. In other embodiments, the concentration of phosphati-dylcholine is about 30–80% of total lipid weight. For example, we are currently examining the PC:cholesterol molar ratios of 7 :3 and 1:1, at both 200- and 400-nm liposome diameters, saturated with calcein self-quenching dyes, at 10 mg/ml total lipid concentrations.

A lipid vesicle may further comprise phosphatidylserine in an amount of about 5–30% of the total lipid weight.

In a separate embodiment, a lipid vesicle comprises sphingomyelin in an amount of 1–30% of the total lipid weight. In other embodiments, the lipid vesicle comprises sphingomyelin in an amount of about 5–30% of the total lipid weight.

Some embodiments include phosphatidylcholine/PE/PS combinations where the phosphatidylcholine/phosphatidylethanolamine is present in an amount of about 5–30% of the total lipid weight.

Lipid vesicles may further comprise cholesterol. Cholesterol is optionally present at an amount relative to the lipid portions of 0.5 cholesterol/lipid or less by molar ratio. In some embodiments, the amount of cholesterol is present at a ratio of 0.01–0.5 cholesterol/lipid or any value or range in between.

It is appreciated that lipid vesicles may contain other materials comprising pro-teins or fragments of proteins that may or may not alter the fluidity of the membrane or provide a membrane with a protein content similar to that of a brain neuron as is known in the art. These include for example integral membrane proteins such as cell surface receptors and transmembrane proteins. It is further appreciated that while the lipid vesicles described are provided as examples, lipid vesicles may com-prise phosphatidylcholine, phosphatidylethanolamine, PS, PI, cholesterol, ceramide, sphingomyelin, and protein in any combination and all combinations are appreciated as envisioned under the invention. In some embodiments, a lipid vesicle includes PC alone to the exclusion of other lipids, cholesterol, or protein.

Methods of forming a lipid vesicle are well known in the art. The experimental platforms will ultimately allow us to determine relationships between mechanical forces and disruption of lipid bilayers and integral membrane proteins. In embodi-ments, the novel sensor will allow us to reduce the complexity of the interaction between insult mechanisms and trauma to the response of individual membrane properties, individual trans-membrane proteins, and extracellular scaffolding prop-erties. This can occur without needing to account for cotemporaneous changes across multiple interacting mechanisms as is the case with whole-cell models. The sensor will also be used to explore and understand the biochemical pathways associated with mTBI. The sensor allows us to isolate and study known integral membrane pro-teins, associate disruption of these proteins and the resultant downstream translation of biomarker proteins, and start to assemble the set of reaction kinetics and network models for each parallel effect. This will then allow for the solution of simultaneous reaction equations to further elucidate the nature of mTBI at the biochemical level.

Methods of forming a lipid vesicle are applicable to the formation of lipid vesicles provided herein as a portion of a biosensor. For example, lipid vesicles formed by techniques for the assembly of the liposome structures are well characterized, since liposomes are the basis for several novel drug delivery systems and therefore well developed. The basic process involves hydration of dry lipid, cholesterol, protein, or other component of the lipid membrane onto a vessel surface from an organic solvent (e.g. chloroform), thereby producing a thin film of dry lipid. This material is then hydrated to solution typically in an aqueous buffer system such as Tris buffered saline, HEPES buffered saline, water, or other suitable buffer known in the art, and forming liposomes as the solution is heated above the liposome phase transition. As many lipids have a phase transition that is below room temperature, heating is not always necessary depending on the total composition of the lipid membrane. Concentric lipid bilayers result, in the form of controlled 30–70 nm diameter liposome spheres. Freeze–thaw processing further refines the liposome morphology. As such, in some embodiments, lipid vesicles are formed by sonication of the hydrated material typically on ice to prevent overheating, or by one or more freeze–thaw cycles.

The resulting liposomes are optionally sized by chromatography or by passing through one or more filters of desired pore size. In some embodiments, a lipid vesicle comprises one or more detection agents. A detection agent is optionally any molecule that is encapsulated by a lipid vesicle that can be released upon vesicle rupture and thereby detected. Illustrative examples of detection, agents include dyes, fluorophores, nucleic acids, proteins, combinations thereof, and the like. By encapsulating one or more detection agents in the liposome (either in the space within the liposome or the lipid mono-layer, bilayer, or multilayer) during the self-assembly process, a lipid vesicle is provided that will allow a detectable color or other change from the trauma-induced liposome disruption that is proportional to the amount of disruption or insult. This affords a very attractive additional feature, whereby a color change readily observed by direct visual observation indicates that a very precise measurement of the subject ought to be the basis for determining prophylaxes and post-trauma expectations. Epidemiological data can also be accumulated rapidly for further assessment of various treatment options to save lives and minimize post-insult conditions.

Illustrative examples of a detection agent include fluorophores such as calcein, pyranine (1-hydroxypyrene-3,6,8-trisulfonic acid), and FAM dye (illustratively 6-carboxyfluorescein). Other fluorophores illustratively include TAMRA AlexaFluor dyes such as AlexaFluor 495 or 590, Cascade Blue, Marina Blue, Pacific Blue, Oregon Green, Rhodamine, Fluorescein, TET, HEX, Cy5, Cy3, Quasar670, and Tetramethylrhodamine. Specific examples of fluorophores include 5- (and -6)-carboxyfluorescein mixed isomers (CF), Alexa Fluor 647 carboxylic acid, succinimidyl ester (Alexa 647) each available from Life Technologies, Grand Island, N.Y.

Other illustrative examples of a detection agent include pH-sensitive dyes. An illustrative example of a pH-sensitive dye is acridine orange and the like.

A detection agent may further comprise a nucleic acid sensitive agent. An illustrative example of a nucleic acid sensitive agent is illustratively Hoechst 33342 (2,5′-bi-1H-benzimidazole, 2′-(4-ethoxyphenyl)-5-(4-methyl-1-piperazinyl)). The casein dyes are similarly nucleic acid sensitive.

In some embodiments, a detection agent is bound to a retention molecule to prevent leakage from the lipid vesicle prior to rupture. Such retention agents are any
membrane inert material such as biotin, polyethylene glycol, antibodies, or other
materials known in the art.

In some embodiments, a detection agent is nucleic acid sensitive. A nucleic acid
is optionally bound to a surface of a housing such as by reaction with a polystyrene
plate. A nucleic acid molecule is optionally animated to promote binding to a polystyrene section of a housing via the amine bond. Illustrative technology is available
from Corning, Inc. Tewksbury, Mass. sold as the DNA-BIND® polystyrene material,
a nucleic acid sensitive dye such as 4′, 6-diamidino-2-phenylindole (DAPI), or the
cell impermeant SYTOX® green nucleic acid stain. When a nucleic acid sensitive
dye is released upon lipid vesicle rupture, it will bind a nucleic acid molecule on the
surface of the housing localizing colorimetric detection of lipid vesicle rupture.

In further embodiments, a detection agent comprises a nucleic acid, antibody,
protein, or other molecule suitable for a specific interaction with a binding partner.
Illustratively, a nucleic acid such as one that is suitable as a primer for a polymerase
chain reaction is used. Extraction of a portion of the extraliposomal solution and its
inclusion in a PCR reaction will positively or negatively discern whether the nucleic
acid based molecule has leaked from the lipid vesicle.

Other detection agents include antibodies that can specifically bind to an antigen
bound to the surface of a housing. The housing surface can then be probed by a
process similar to an enzyme-linked immunosorbent assay (ELISA) to detect the
presence or absence of the antibody. It is appreciated that other detection agents are
similarly suitable.

Freeze–thaw processing during lipid vesicle formation (e.g. 2–3 cycles) further
refines the liposome morphology and provides for encapsulation of one or more
detection agents from a biphasic mixture for colorimetric sensor features. Note that
certain fluorophores are activated upon exposure to solvent reagents and or water,
thereby effecting the color change for direct insult observation. As such in some
embodiments, one or more fluorophores are encapsulated in a lipid vesicle formed
in a nonaqueous solution. The resulting vesicles are then washed and placed in an
aqueous medium for association with housing. Upon rupture due to an event force,
the detection agent escapes the lipid vesicle and is detectable by a direct color change.
This provides rapid and easy identification of a situation requiring medical attention
to the wearer of the biosensor.

The liposome-based sensor can be incorporated in a variety of ATMs and instrumented head forms (collectively housing) currently available, supplanting, or at least
supplementing the less effective pressure sensors and accelerometers that previously
provided far less than optimal correlation to mild traumatic brain injury or traumatic
brain injury at best. As the applications are validated for the sensor, a variety of
housing options are available. These include lamellar encapsulation of the liposomes
themselves, incorporation in solution, and containment in honey-comb interlaced
sheets of material as well as tablets or ampoules for convenient attachment at vulnerability points of interest. Arrays of the sensors can also be implemented to provide
dimensionality and spatial control of the event characterization. Placement of the

sensor as close as possible to the vulnerable areas of concern thus provides a direct and more accurate measure of the tissue susceptibility to a threat combination or singular test scenario.

Housing is either a surface whereby the lipid vesicles are exposed directly to the environment, or the housing encapsulates the lipid vesicles whereby the housing has at least one surface that will transmit an event force from the external environment to the lipid vesicle. In an embodiment, at least one portion of a housing is transparent to a wavelength of light emitted or reflected by a detection agent. In another embodiment, a housing is in a capsule form. In a further embodiment, a housing is in a cubic or rectangular prism form. In yet another embodiment, a housing is in a spherical, sheet, curvilinear, or other two-dimensional or three-dimensional shape. One example of a suitable housing is similar to a dialysis cassette. In embodiments such a housing has a membrane on one or both sides that will transmit an event force to a lipid vesicle contained within the housing. Such embodiments also provide the ability to encapsulate one or more detection agents into the vesicle and then transfer buffer or wash away any excess dye after vesicle formation simply by buffer exchange right in the housing.

Optionally, a housing is in the form of a capsule. Capsules can be formed of any material traditionally known in the art that will transmit an event force to a lipid vesicle. Illustrative materials comprise gelatin, starch, casein, chitosan, soya bean protein, safflower protein, alginates, gellan gum, carrageenan, xanthan gum, phthalated gelatin, succinated gelatin, cellulose phthalate-acetate, oleoresin, polyvinyl acetate, hydroxypropyl methyl cellulose, polymerisates of acrylic or methacrylic esters, polyvinyl acetate-phthalate and combinations thereof. A housing is optionally formed in whole or in part of polymeric materials. Illustrative examples include flexible vinyls (e.g. polyvinylchloride), polyamides, polypropylene, norell, polysulfone, ABS, polyethylene, natural and synthetic rubbers, among many others.

The molecular biosensors provided have the capability to detect, measure, quantify, and optionally correlate an event force to the likelihood or severity of traumatic brain injury or mild traumatic brain injury suffered by a wearer of the biosensor. As such, processes of detecting and, in some embodiments, quantifying blast or other direct event force using self-assembled liposome structures as a unique sensor are provided. The liposome structures are configured and packaged in a manner where the sensors can be affixed to a subject's clothing, a helmet, body armor, or personal protective equipment in a manner that provides direct indication of the trauma received at the point of attachment. As such, in some embodiments, a molecular biosensor is affixed to an item of clothing or protective equipment such as a helmet or to a traditional clothing form. A molecular biosensor is optionally affixed to a building or vehicle surface. A molecular biosensor has utility for the detection of blunt and ballistic trauma, as well as the convoled effect of shock waves associated with blast trauma, received by the body tissues of a subject during equivalent events. The sensors provided represent the first real and direct measure by which insult is correlated to injury. The disruption of the phospholipid bilayer in human tissue resulting from such forces is directly measured by the liposome sensors in the most meaningful way possible—by the identical disruption that occurs to a subject.

An apparatus and process of detecting a traumatic event is provided including subjecting a molecular biosensor to an event force, and analyzing the biosensor or lipid vesicle portion for alteration indicative of an event force sufficient to produce traumatic brain injury or mild traumatic brain injury in a subject subsequent to the initial trauma. A shock wave generator is one possible source of an event force. A blunt force trauma is optionally produced by any source of such force. Experimentally blunt force events are produced by fluid percussion, cortical impact, and weight drop/impact acceleration sources.

A molecular biosensor is used in a process of detecting a traumatic event, or in the absence of a lipid vesicle, a housing is used. The presence or absence of a traumatic event is determined by an alteration in the lipid vesicle itself or by an alteration in the amount, type, binding, or other characteristic of a detection agent present in the lipid vesicle or on the lipid vesicle.

In embodiments, a traumatic event is determined by an alteration in the molecular structure or orientation of one or more molecules that make up a lipid vesicle. Illustratively, circular dichroism is used to detect molecular alterations in one or more components of a lipid vesicle. The sample material to be analyzed is contained in a quartz cylinder, within which are spacers to accommodate smaller sample vessels. An alternate sample container comprises a Hellma Analytics photometric micro tray cell cap. Alternate packaging schemes may be developed to provide insult maps across curved and rectilinear tessellations for certain applications. Linearly polarized light is passed through the analyte. In a chiral material such as the liposome, the right and left circularly polarized components travel at different velocities and are differentially absorbed. This results in the light exiting the analyte with elliptical polarization, and the analyte is deemed to possess circular dichroism (CD). The magnitude of CD is expressed as the molecular ellipticity θ:

$$\Theta = 4500/\pi(\epsilon_L - \epsilon_R)\log_e 10 \tag{1}$$

where ϵ_L and ϵ_R represent the molecular extinction coefficients for the left and right circularly polarized light beam components, respectively. The difference between the extinction coefficients is

$$\Delta\epsilon = (\epsilon_L - \epsilon_R) = 1/LC \log_{10}(I_R/I_L) \tag{2}$$

where L is the absorbing layer thickness (cm), C is the molar concentration, I_R and I_L are the intensities of the right and left circularly polarized light beams, respectively, after passing through the analyte. Θ then becomes

$$\Theta = 4500/\pi LC \ 10 \log_{10}(I_R/I_L) \tag{3}$$

Circular dichroism spectrometers such as the J-815 from Jasco Corporation, Easton, MD, measure CD changes of the magnitude produced in the sensor. Differential CD is created by a disruption of the intramolecular chiral interactions of three-dimensional molecular structures as well as the additional chiral symmetry breaking of nonchiral molecules in the sensor material construction. Measurements in changes of CD are indicative of alterations in the structure of the lipid vesicle and indicative of an event force.

In embodiments, a traumatic event is detected by the analysis of a medium external to a lipid vesicle. For example, and in embodiments, a lipid vesicle includes one or more detection agents within the vesicle. The leakage of a detection agent(s) into the extraliposomal space is indicative of damage to the lipid membrane such as rupture or more minor damage. Optionally, a detection agent is a fluorophore. In embodiments, the fluorescence is quenched due to high dye concentration internal to the liposome, and unquenched upon release of the dye molecules into the surrounding solvent upon event force. Instrumented techniques based on atomic force microscopy, confocal fluorescence microscopy, and fluorescence recovery after photo bleaching (FRAP) optionally coupled with colorimetric fluorometry (detection of light intensity based on leakage of dye from the disrupted cell wall) are used to analyze liposome disruption and failure criteria. The rate of detection agent release is optionally accomplished through cholesterol-influenced bilayer properties where higher levels of cholesterol typically equate to less fluid and more event force resistant membrane.

The molecular biosensors and processes provided find usefulness in many arenas such as in military applications for the design of protective equipment such as body armor and helmets; rapid diagnosis or prediction of possible traumatic brain injury or mild traumatic brain injury in a subject receiving an event force in the field to provide or indicate the need for medical intervention; as a point sensor or array of sensors to provide 2D mapping of trauma; as a sensor useful in the design and use of sport-related protective headgear in which concussions and other brain injuries are a concern; as a sensor for research, testing and/or development of protective equipment by athletic equipment manufacturers and military equipment manufacturers; and as a research tool for the understanding of the molecular results of forces that produce traumatic events and complications such as traumatic brain injury or mild traumatic brain injury.

Various modifications of the present invention, in addition to those shown and described herein, will be apparent to these skilled in the art of the above description. Such modifications are also intended to fall within the scope of the appended claims.

Patents and publications mentioned in the specification are indicative of the levels of those skilled in the art to which the invention pertains. These patents and publications are incorporated herein by reference to the same extent as if each individual application or publication was specifically and individually incorporated herein by reference.

The foregoing description is illustrative of particular embodiments of the invention, but is not meant to be a limitation upon the practice thereof.

The invention claimed is:

1. A molecular biosensor for determining blast, ballistic, and blunt trauma comprising:
   a lipid vesicle comprising a detection agent;
   a housing, said vesicle contained on or within said housing, said housing having a portion capable of transmitting a blast, ballistic, or blunt traumatic force generated external to said housing to said vesicle;

wherein said vesicle comprises phosphatidylcholine, phosphatidylserine, phosphatidylethanolamine, phosphatidylinositol, sphingomyelin cholesterol, ceramide, or combinations thereof;

wherein the phosphatidylcholine, phosphatidylserine, phosphatidylethanolamine, phosphatidylinositol, sphingomyelin cholesterol, ceramide, or combinations

thereof are inserted into a circular dichroism (CD) meter to measure a differential absorbance of left and right circularly polarized light; and

further wherein a differential circular dichroism is created by disruption of intramolecular chiral iterations of three-dimensional molecular structures as well as a chiral symmetry breaking of nonchiral molecules in a sensor material construction.

2. The biosensor of claim **1** wherein said vesicle consists of phosphatidylcholine.

3. The biosensor of claim **1** wherein said vesicle comprises greater than 50% total lipid of phosphatidylcholine, phosphatidylethanolamine, or a combination thereof, and

wherein said housing is in the form of a lamellar structure, interlaced sheet, capsule, or tablet.

4. The biosensor of claim **1** wherein said housing is in the form of a capsule.

5. The biosensor of claim **1** wherein said vesicle further comprises a detection agent.

6. The biosensor of claim **1** wherein said detection agent consists of a fluorophore.

* * * * *

Appendix 2
Significance of the NMDA Cell Surface Receptor

Significant modeling efforts included an assessment of the *N*-methyl-*D*-aspartate (NMDA) integral membrane protein. The proposed analysis was to elucidate the mechanisms by which trauma-induced disruption of the NMDA leads to nucleation of neuropathologies associated with mTBI. In particular, the work of the Armstead group at University of Pennsylvania (Armstead et al., 2011) is of great interest, as they elucidate putative mechanisms for how serine protease tPA impairs the increase of the internal diameter of blood vessels following fluid percussion brain injury by mechanisms correlated to mitogen-activated protein kinase (MAPK) isoforms. Armstead addresses specific aspects of mild traumatic brain injury and signal pathway effects/defects, especially those associated with MAPK.

Part of the motivation for the research is that no objective biologic or physiologic diagnostic measures exist for mTBI, despite its significance as the signature wound in Iraq and Afghanistan. Upon successful development of suitable field-based sensor systems for detecting mTBI, improved understanding of neurobiochemical signal cascades proximal to trauma is required to develop efficacious therapeutics. Of great interest in the study of the response to TBI is the MAPK signaling pathway, activated through phosphorylation in response to activating signals (mitogens) occurring as a result of the traumatic event. Results lend themselves to ancillary issues of distinguishing TBI from post-traumatic stress syndromes and other psychiatric conditions, and improved basis for cataloging epidemiology's, including those in the infant category.

Following the Armstead analysis, TBI results in an "uncoupling" of blood flow and metabolism, evidenced as either insufficient blood flow to the brain (ischemia) or increased blood flow and diffuse swelling (hyperemia). It is the former (cerebral hypoperfusion) that is most frequently observed in cases of TBI. Since the mechanism(s) for TBI mediation in the brain are not clearly established, potentially interrelated studies are of great importance in defining the current work to explore mediation of cerebral hemodynamics following TBI.

Glutamate binds ionotropic receptors, causing cerebrovasodilation, providing a potential means for coupling of local metabolism and blood flow. Indeed, excessive glutamatergic activation of NMDA causes tissue damage through cell death, while NMDA receptor antagonists may protect against TBI. While cerebral hemodynamics implicate with NMDA activity, and NMDA-induced vasodilation reverses to vasoconstriction after TBI (suggesting the NMDA antagonist treatment)—the NMDA mechanism is not well understood.

We've seen in previous work that EEIIMD (a tPA antagonist—see terms at the end of this section) inhibits tPA mediation of vascular action (e.g., tPA mediated NMDA hypoperfusion). It is also observed that there is elevated tPA in the CSF after FPI, and more so in newborn piglets than infants. So, we conclude here that EEIIMD prevents the NMDA mediation of dilation to vasoconstriction following TBI. But the mechanism for tPA hemodynamic impairment is not understood. However, we do know that tPA is mediated by the MAPKs, which are up-regulated by TBI. We also know that ERK activation results in hypoperfusion, concurrent with reduced NMDA dilation after TBI. How p38 and ERK MAPKs modulate tPA regulation of NMDA is not known, but we hypothesize as stated in our initial introduction:

> Serine protease tPA impairs increase of internal diameter of blood vessels following fluid percussion brain injury by a mechanism correlated to a mitogen activated protein kinase (MAPK) isoform.

The experiments performed at the University of Pennsylvania used the following:

- Closed cranial window, artificial CSF
- Fluid percussion injury, pendulum piston impactor
- 11-group experimental protocol
- ELISA quantification of CSF ERK, p38, and JNK phosphorylations
- Statistical analysis of pial artery diameter and cerebral spinal fluid

Justification for the use of newborn piglets was based on their gyrencephalic brains with significant white matter ratio similar to humans. Their convenient size readily affords hemodynamic investigations; and newborns exhibit increased sensitivity to FPI/TBI, regarding hypoperfusion, concurrent decreased pial artery diameter, and impaired ability to increase the needed blood flow (e.g., impaired vasodilator response).

The piglets were exposed to the dura, and remaining matter was cut and retracted. A glass cover slip was placed over the cranial window. The window volume was accessed via three hypodermic ports fixtured in a retaining ring. Pial artery diameters were measured microscopically, using videography. The window was filled with CSF similar to endogenous conditions:

> Millimolar concentrations: 3.0 KCl, 1.5 $MgCl_2$, 1.5 $CaCl_2$, 132 NaCl, 6.6 urea, 3.7 dextrose, 24.6 $NaHCO_3$.

Fluid percussion brain injury (FPI) was produced through the craniectomy by pressure impulse insults to the intact dura. In the FPI device from AmScien Instruments, based on the original design from the VA Medical College, energy was transferred via pneumatic pressure. In the original Armstead experiments described herein, the

FPI piston was struck with a 4.8 kg pendulum. The pendulum drop height was controlled to produce FPI intensity of 1.9–2.3 atm (atmospheric pressure units), monitored via an optical sensor trigger and oscilloscope. The amplitude of the pressure pulse was used as the metric for FPI. The Armstead experimental protocol was the following:

- Small arteries (120–160 μm resting diameter) and
- Arterioles (50–70 μm) utilized
- 2–3 mL artificial CSF flushing for 30 seconds
- 300 μL/500 μL total window fluid volume sampling
- Vehicle was 0.9% saline for all groups, except 100 μL dimethyl sulfoxide diluted with 9.9 mL of 0.9% saline for 0126 and SP600125
- Eleven experimental groups ($n=5$): phosphorylation of JNK, ERK, and MAPK tested initially and after 1 hour following FPI, for #1–7

1. Sham control with vehicle	7. FPI with tPA and D-JNKI1
2. FPI with vehicle (artificial CSI)	8. FPI with U0126
3. FPI with tPA	9. FPI with SB203580
4. FPI with tPA and SB203580	10. FPI with SP600125
5. FPI with tPA and U0126	11. FPI with D-JNKI1
6. FPI with plasminogen activator and SP600125	

- For sham control and FPI experiments, vascular response (% change in artery diameter) was tested for agonists NMDA, glutamate, and papaverine initially and after 1 hour following FPI in #1–7 above
- #8–11 provided ability to normalize results of #4–7

ELISA assays for phosphorylated MAPK isoforms were normalized to total form and expressed as percentage of control. Pial artery diameters and MAPK values in CSF were analyzed with ANOVA for repeated measures. For significant values (p less than 0.05, for probability of seeing that value or difference, given the null hypothesis): Fisher protected least significant difference test applied (a t-test for *all* pairwise comparisons in the ANOVA). Values were then reported as the sample mean, ± the standard error of the mean (standard deviation).

- Bar chart extensions in the following figures are represented by
- *$P<0.05$ compared with control
- +$P<0.05$ compared with FPI with vehicle
- #$P<0.05$ compared with FPI and tPAI

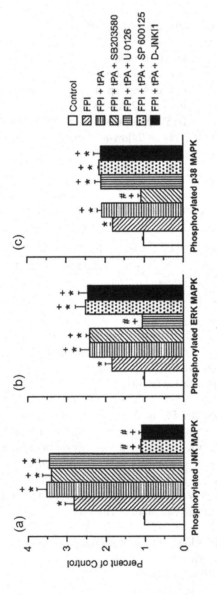

Adapted from (Armstead et al., 2011) under Open Access BMC license agreement

- JNK, ERK, and p38 MAPK isoforms up-regulated (↑) after FPI in the proportion JNK>ERK≈p38
- tPA (10-7M) augmented the increases after FPI
- JNK antagonists SP600125 and D-JNKI1 blocked JNK ↑ after FPI, but left ERK and p38 unchanged
- ERK antagonist U0126 blocked ERK ↑ after FPI, but left JNK and p38 unchanged
- P38 antagonist SB203580 blocked p38 ↑, but left JNK and ERK unchanged
- Data not shown also indicated tPA having no effect on CSF MAPK without the FP

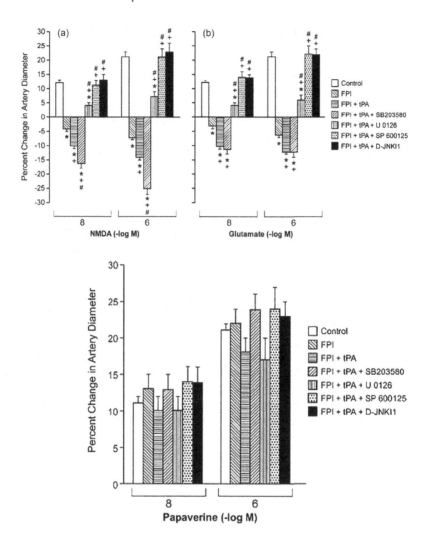

Adapted from (Armstead et al., 2011) under Open Access BMC license agreement

- Data not shown demonstrate artery/arteriole dilation in controls
- As seen in these two data sets, FPI resulted in reversal of dilation to vasoconstriction for NMDA and glutamate, but not for papaverine; and tPA pretreatment is seen to enhance these trends (except for papaverine)
- tPA with JNK antagonists SP600125 and D-JNKI1 blocked NMDA and glutamate-induced vasoconstriction; likewise with U 0126- but to a lesser extent
- However, p38 inhibitor SB 203580 with tPA augmented vasoconstriction, but ONLY when tPA was administered
- Papaverine-induced vasodilation was unaffected by FPI or the co-administered treatments

DISCUSSION, CONCLUSIONS, AND PROPOSED FUTURE WORK

We've seen that JNK, ERK, and p38 MAPK isoforms are up-regulated (↑) after FPI in the proportion JNK>ERK≈p38. Also, JNK, ERK, and p38 MAPK isoforms were blocked by their respective antagonists, and unchanged by other MAPK inhibitors—demonstrating the separability of the treatments. Dilation was reversed to vasoconstriction after FPI, and constriction was enhanced by tPA with glutamate and NMDA; and reversed by antagonists. For this reason, we can confidently address the practicability of modulation of tPA, MAPK, and NMDA receptors (in drawing conclusions to our hypothesis).

NMDA → vasodilation; typically mediated by tPA

FPI → JNK↑

FPI → tPA↑; and therefore FPI → impaired dilation

Therefore: *JNK* performs a key function in up regulation of the tPA mediation (impairment) of NMDA-induced vasodilation after FPI. This is a restatement of our original hypothesis:

> serine protease tPA impairs increase of internal diameter of blood vessels following fluid percussion brain injury by a mechanism correlated to a mitogen activated protein kinase (MAPK) isoform

Since tPA did NOT affect NMDA without FPI; nor did it affect papaverine dilation, tPA interacts with NMDA in a manner influenced by the effects of FPI. tPA exhibits a selective effect on the excitatory amino acids. Since tPA itself is a tissue-type plasminogen activator, we would expect its effect with additional activators to enhance—rather than the observed detractor effect. While the biochemical basis for this is not yet understood—the effect certainly appears to be activated by the MAPK signaling pathways.

While we know from previous work that CSF analysis directly reflects parenchymal activity following FPI—the closed cranial window approach does not provide cellular site specificity—such that activity is not differentiated for neurons, glia, vascular smooth muscle, and endothelial sources. Nonetheless, the technique employed in this work represents the most effective analytical system available at this time.

Next, we would explore the links between tPA and glutamate receptors, and their relationship to cell death, nitric oxide synthesis, and toxicity causation for neuronal cell loss, since it is known that NMDA receptor activation→dilation, and relates to local metabolism and blood flow. This could lead to a greatly improved understanding of the means by which we could therapeutically balance cerebral effects of ischemia or hyperemia, creating optimal tissue response to FPI.

We would also explore and identify the specific vasoconstrictor likely produced (endothelin?) after FPI in concert with NMDA activation and vasoconstriction, to account for the degree of constriction observed; and to search for explanations that account for dilator loss in such a mechanism as well. Since the impairment of NMDA function results in the disruption of optimal hemodynamic regulation following FPI, the work accomplished thus far provides a basis for continued exploration into blood flow dynamics and prophylaxis, both pre- and post-trauma.

It may well be that a viable treatment is to impair the hypoperfusion effects of tPA activation of MAPK signaling, post-trauma, providing and enhancing proper balance of blood flow factors and subsequent damage repair or tissue regeneration mechanisms. Coupled to this potential treatment is our observation that the p38 MAPK may be advantageous—in contrast to the effects of ERK or JNK—so that up regulation of p38 MAPK pathways may prove efficacious.

A potential plan forward to further explore and extend the work of Armstead is to identify and quantify any thresholds associated with the observed FPI effects. We would also

- Correlate FPI intensity levels to the dynamic range of anticipated TBI events to include ballistic and blast overpressure traumas
- Explore the relationship between the peak pressure metric employed in the current work to the impulse function (the time integration of the applied force) rather than relying upon simply the peak applied value
- Identify the relationship between white matter ratio, tissue density, and alternate animal models for correlation to human injury effects
- Finally, upon determination of thresholds for injury mechanisms of interest, correlate observable hemodynamic metrics with sensor construct data sets external to the skull to provide/develop well-correlated predictive indicators

Terms and acronyms for ready reference in this section

1. NMDA: N-methyl-D-aspartate
2. PAD: pial artery dilation; pia: inner layer of the meninges
3. FPI: fluid percussion brain injury
4. tPA: tissue type plasminogen activator; a serine protease; converts plasminogen to active protease plasmin; observed in the cerebral spinal fluid in typically 10-7 M after FPI
5. EEIIMD: a tPA antagonist; peptide derived from endogenous plasminogen activator inhibitor-1
6. MAPK: mitogen-activated protein kinase
7. ERK, p38, JNK: members of the MAPK family
8. JNK: c-Jun N-terminal kinase
9. TBI: traumatic brain injury
10. CSF: cerebrospinal fluid
11. SP600125, D-JNKI1: JNK antagonists
12. U 0126: ERK antagonist
13. SB203580: p38 inhibitor
14. Glutamate: an excitatory amino acid transmitter in brain
15. NMDA, kainate, alpha-amino-3-hydroxy-5-methyl-4-isoxazazolepropionate: ionotropic receptor subtypes
16. ANOVA: analysis of variance; a collection of statistical models
17. Papaverine: a vasodilator; improves blood flow
18. ELISA: enzyme-linked immunosorbent assay

REFERENCE

Armstead, W.M., J.W. Kiessling, J. Riley, D.B. Cines, and A.A. Higazi. 2011. tPA contributes to impaired NMDA cerebrovasodilation after traumatic brain injury through activation of JNK MAPK. *Neurol Res.* 33(7):726–733. Accessed March 23, 2020 at https://www.ncbi.nlm.nih.gov/pmc/articles/PMC3523283/

Appendix 3
Neuroproteomics, Protein Folding, Transcription Factors, and Epigenetics for TBI Research

We first examine the characteristics of the eukaryotic chromosome, including anchoring proteins, coiling, and associated methods of compaction. Specific proteins anchor to the plasma membrane and interact with DNA during the supercoiling process. They also function to hold together loop domain centers. More complex arrangements exhibit in eukaryotic chromosome protein anchors, where the histone amino acids bind DNA, changing conformation and effecting the compaction process. Nuclear matrix proteins bond to chromosome 30 nm fibers for compaction through radial loop domains; and during mitosis, chromosomes are anchored to the spindle—a process unique to the eukaryote.

It is instructive to consider the varied roles of the structural maintenance family (SMC)—in particular, the cohesin–protein complex. In addition to its "traditional" roles of controlling sister chromatid separation during metaphase, facilitating chromosome spindle attachment, and assisting in DNA repair, cohesin provides a stabilizing role in insulator effects on the major histocompatibility complex (MHC) transcription, through interactions with the MHC insulators. It is thought that the cohesin ring structure is able to encircle sister chromatids, providing structure and support for uncondensed sister chromatid coherence, condensation, and freeing of chromatids at prophase, and separation of condensed chromatids at anaphase.

Chromosome assembly and segregation in eukaryotes is further assisted by another SMC family member: the large protein complex condensin that travels from the cytoplasm to the nucleus during the start of the M phase. Condensin also binds to chromosomes and assists in compacting radial loops. Two distinct SMC family condensing complexes are also seen to assist with the assembly of condensed chromosomes.

Chromosomes must be coiled for the cell structures to accommodate their size. Coiling may involve circular, closed loop, naked (that is, no histone proteins), double-stranded DNA molecules. These need to be coiled greater than 1,000 fold for sufficient compaction into the cell. Sister chromatids at the conclusion of interphase are double helix DNA, wrapped around histones, forming the nucleosome. This structure is supercoiled into 10- and 30-nm beads. The eukaryotic chromosomes are fairly linear (relative to bacterial chromosomes), and coil into roughly an "X" shape. Seven-fold compacted, "beads on a string" structures are further compacted via histone mediation to 30-nm fibers, getting us to 50-fold compaction.

The next sequence of compaction utilizes nuclear matrix proteins wound around histones, forming coiled beads or nucleosomes. Further twisting into a 700-nm-wide solenoidal structure is accomplished, and radial loop domains provide additional functionality. We call this DNA/histone combination "chromatin." Further compaction occurs in the heterochromatin. In bacterial chromosomes, DNA binding proteins bind loop domains that then coil in a 10-fold compaction. Then the loop domain structures are "supercoiled" by enzyme catalyzed twisting and compaction of the DNA into itself. After looping and supercoiling, the chromosome then fits into the nucleoid.

Protein domain structure relates to the binding of proteins to DNA, binding of proteins to other proteins, and to domains that activate transcription. We define protein domains as portions of protein sequences with independent function and the ability to fold and reconfigure in an independent manner. One or more motifs comprise the domain, with each motif providing specific structure to the domain. Domains that are part of a macromolecular structure modulate molecular flexibility through conformational changes that serve to alter the accessibility of certain regions, alter reactivity, and positioning—all functioning as a complex lock and key mechanism. We can further analogize domains as Swiss army knives, with varieties of interaction sequences that can be effected in sequencing and activating docking sites, functionalities, and temporal relationships. We suggest as well a thorough examination of the actions of hydrogen bonded water networks that mediate domains and the thermodynamic "Anfinsen" partition function within the cellular confines—where much of the domain activity occurs.

Specific relevance to our three categories of DNA binding, proteinprotein binding, and transcription activation is evidenced in the wealth of evolving knowledge and validation through X-ray diffraction, especially crystallization of metastable domains—and emerging methodologies such as neutron spin echo spectroscopy, and protein tomography, providing visualization of individual domains.

The major groove in a DNA double helix accommodates binding proteins with various domain modules. Dimeric binding proteins provide dimeric receptors and many permutations at a given DNA binding site (in fact there are currently more than 300 complexes and bioinformatic structural predictions on the protein database, along with about 45,000 3D structures confirmed experimentally thus far.) Conserved protein domains provide binding to specific activators controlling signal transduction pathways. A key example of this is the zinc finger domain, exhibiting conformational changes when bound to DNA binding sites.

Several domains can be bound together resulting in further diversity of physicochemical properties and multifunctionality—not just to provide binding sites and regulatory functions—but also as building blocks for biological assemblies and tissue structure. Some excellent examples are binding of DNA helicase to DNA Pol III holoenzyme, regulating DNA strand separation; initiation of eukaryotic DNA replication via the origin recognition protein complex; and finally, the glycolytic enzyme pyruvate kinase, containing a beta regulatory domain, alpha/beta substrate binding domain, and an alpha/beta nucleotide binding domain, all connected with polypeptide linkers.

Protein domains that activate or regulate transcription are exemplified by the leucine zipper domain, mediating DNA binding, dimerization, and transcription initiation. Zinc finger domains (ZnF_GATA) are transcription factors that bind promoter sequences—as a further example of transcription activation domains. In general, we see activators binding enhancers, and repressors binding silencers. Domain dimerization for transcription includes the homodimer and heterodimer formation of identical transcription factors, or different factors, respectively. TFIID and Mediator regulatory transcription factors communicate directly with RNA polymerase, and indirect transcription regulation involves recruiting nucleosome positioning proteins, as well as histone and DNA modifying proteins in the chromatin-remodeling process.

Transcription factors are influenced by two categories of signals: first are external signals including cytokines, growth factors, hormones, etc., along with stress type responses; and second is a set of intrinsic factors to include transcription, DNA replication, and chromosome modification/segregation. Histone modifications result in repression or alteration of genome function (chromatin deregulation) and, along with extrinsic factor regulation, collectively actualize the epigenetic condition determining modifications and portions of the genetic sequence to be available for transcription. Post-translational modification of histones (HPTMs) includes acetylation, phosphorylation, and methylation, as well as ubiquitylation and sumoylation—the latter two via much larger peptides.

DNA methylation represents a potentially informative form of epigenetic modulation of transcription. Three models describe how these modifications activate or repress transcription: (1) HPTM alters chromatin structure (e.g. by changing electronic charge); (2) HPTM prevents binding of a negative factor to the chromatin template; (3) HPTM creates a binding site for a positive factor. One is typically a *cis* modifier; and the other two are usually *trans* modifiers.

Acetylation (adding an acyl to an active hydrogen site [H_3N^+] on an N tail of lysine residues) activates transcription at H3-K9, K14, K18, K56; H4-K5, K8, K12, K16; H2A, and H2B-K6, K7, K16, and K17. Transcription is also activated by phosphorylation at H3-S10; arginine methylation at H3-K4, K36, and K79; and ubiquitylation at H2B-K123. Transcription is repressed via arginine methylation at H3-K9, K27, and H4-K20; ubiquitylation at H2A-K119; and sumoylation at H3, H4-K5, K8, K12, K16, H2A-K126, and H2B-K6, K7, K16, and K17. Acetylation is catalyzed by site-specific histone acetyltransferases and reversed with deacetylase (HDAC). Typically, activators recruit HATs and repressors recruit HDACs. This compacts the nucleosome (model 1) by converting a positive charge at the H_3N^+ to a partial negative charge due to the =O bond that repels the DNA and tail-opening up binding sites on the DNA as well as possibly decompacting the nucleosome, making chromatin more easily activated.

Via model 3, bromodomains bind to acetylated lysine residues, and may be part of HAT motifs such as CPB/p300 and Gcn5, in remodeling complexes like Swi/Snf, which promote binding to chromatin. Other bromodomains may also be attracted such as Taf1 and TFIID complexes, Rsc4 in Rsc, and Brd in various proteins. Transcription repression results from deacetylation via HDAC Sir2 enzymes, acting with cofactors such as Rpd3 in the HDAC Sin3 complex, or Rpd3 in H3K36me. This suppresses RNA Pol II, and regulates the steps of the transcription cycle.

Phosphorylation location is also determined by enzymatic specificity. Proposed mechanisms involve all three models: residue cluster phosphorylation in H1 alters DNA binding affinity, locally increasing transcription potential of the chromatin; ala model 2: HP1 binding affinity is lowered throughout mitosis via phosphorylation at H3-S10; and ala model 3: H3-S10 phosphorylation is recognized by an adapter protein, inducing transcription. Models 1 and 3 apply to histone lysine methylation, depending on the site. One proposed mechanism is SUV39H1/Clr4 signaling to recruit HP1 or Swi6 and Chp2, effecting replacement of H3 with CENP-A. Demethylation is accomplished via deimination, removing methyls, or via demethylase. Arginine methylation occurs mostly in the nucleus, positively or negatively affecting transcription via methyltransferases that vary in substrate specificity, cellular localization, and targeting. Deimination can activate or repress transcription, depending on tissue and location. Finally, ubiquitylation and sumoylation can modulate transcription by activating methylation or regulating acetylation.

DNA methylation prevents binding of transcription factors to promoters. It is thought that MeCP2 recruits Sin3A HDAC and HKMT to methylated sites (marks), repressing transcription initiation complexes. Gene expression can be activated via mediation of DNA methylation by MDB proteins at CpG islands, which can repress or potentially activate transcription at the CpG islands (an example of RNA activation). In a different example, methylation outside the islands can modulate stem cell differentiation for blood cell type. Imprinting can also result, with parent sex specific positive transcription activation via variable epigenetic cluster modification at combinations of sites, providing transcription activation variants. RISC and RITS processes modulate genetic expression (activate and/or repress) during various phases of the cell cycle.

Comparing repressive effects of chromatin and DNA modifications to those of RNAi, we first provide a basic overview of RISC post-translational silencing and RITS pre-translational silencing. RISC is the RNA silence-inducing complex, directed by siRNAs and miRNAs, plus Slicer; and RITS is the RNA-induced transcriptional silencing complex, which uses siRNA base pairing, stabilized by methylated H3, providing epigenetic chromatin modification. In RISC, siRNA base pairs to mRNAs, recognizing them for degradation and cleavage; while in RITS, the siRNAs target the silencing complex to regions of the chromosome designated for chromatin modification.

RNAi pathways involve enzymatic processing of primary miRNA or dsRNA. Similar to the recruitment via histone modifications of transferases to modified histones, RNAi recruits enzymes to the chromatin that provide modifications. For miRNA, Drosha generates pre-miRNA that is transported to the nucleus by exportin5—while the Dicer cleaves the dsRNA in the cytosol. Dicer removes the stem loop used in miRNA, producing the mature miRNA complex that gets incorporated in the RISC complex after helicase cleavage; while helicase also cleaves the siRNA duplexes for their subsequent incorporation into the RISC. The pathways for miRNA and siRNA to their RISC complexes—along with Argonaute (a common feature of RITS and RISC)—provide a means to target and cleave mRNA in the RISC complex complementary to the RNAi. But in the RITS silencing complex, the target DNA is silenced via the formation of heterochromatin. Argonaute binds the guide strand to the direct silencing of genes.

RISC thereby controls the gene silencing process—which can include the chromatin modifications discussed. RNAi mechanisms that are unique include the immune defense against viruses and transposons—roles performed by RNAi in addition to gene regulation. Demethylation and remethylation during cell differentiation and development temporarily program epigenetic patterns, a more permanent effect than the DNMT catalyzed methylation patterns in non-CpG DNA. RITS controls histone modification (and formation of heterochromatin) prior to transcription. Additional unique roles of RNAi include control of morphogenesis, stem cell maintenance—especially during differentiation and a role as either oncogenes or tumor suppressors. Plural dynamic modifications result in the specificity of position, timing, and systematic control of gene transcription through combinations of the processes discussed.

RNAi can serve to destroy mRNA in addition to modulating translation—another unique process. It can also silence a promoter, in addition to directing chromatin modification. And unlike a simple and direct histone modification that affects more than just the chromosome and branch location of interest—RNAi can be used to switch on and off a specific site of interest—making it quite valuable for manipulation of gene expression. RNAi silencing represents the more efficient of the processes, owing to this specificity and precision of its processes, compared with our menu of histone modifications. Control via RNAi of heterochromatin initiation and assembly requires the process pathways described for RITS, where it is postulated that specific nuclear regions are assembled with specific configurations and timing directed via the RNAi RITS pathways—in which RITS recruits the proper enzymes (such as RDRC) to select RNA templates.

As the human epigenome catalog is established, the direct role of RNAi for epigenetic modulation, gene silencing, chromatin remodeling, cell development, immune responses, and other interrelated roles of the processes discussed will be elucidated. It is anticipated this will lead to an improved understanding of a wide range of interrelated and cascaded pathologies, along with their practicable cognate therapeutic protocols.

Tissue microarray analyses provide links between global histone modifications (such as acetylation and methylation of lysine and arginine residues) and susceptibility to oncogene promotion and recurrence of prostate cancer. Mechanisms for this may include chromosomal translocation enhancement via modifications of exchange kinetics (by modulating the accessibility, reactivity, and positioning of protein domains that activate transcription) between nonhomologous chromosomes, producing constitutively active chimeric proteins or oncogenic receptor activation, and inappropriate expression of genes regulating growth and amplification of protooncogenic segments—along with angiogenesis increase, avoidance of apoptosis, diminution of DNA repair capability, and increased tissue invasion capacity. DNA microarray analysis of cancer phenotypes indicates signaling pathway aberrations effected via the disruption of intracellular antagonists, receptor blockers, reconfigured enhanceosomes, and spatiotemporal signal chains, some of which result in oncogenic progression. Changes in gene expression patterns resulting from chromatin modifications manifest, for example, as various lymphomas that are otherwise phenotypically similar. Additional examples are loss of function mutation in Swi/Snf, increasing proliferation of E2F and leading to cancer progression, and activation of prometastatic genes such as urokinase plasminogen activator and heparanase.

We've explored the conserved role of intragenic differential methylation across the gene body, at the 5' promoter, in the introns (intragenic), at the 3' end of the DNA transcript, and in the gene-sparse intergenic region, and where the C-phosphate-G (CG) nucleotides occur in higher concentrations or islands (CpGs). Tissue-specific methylation via DNMT enzymes indeed regulates promoter activity in the intragenic and 3' transcript end, and in intergenic regions; regulating alternative (distant) promoters (also termed alternative CGI promoters); underscoring the importance of histone methylation and acetylation (along with phosphorylation, ubiquitination, ADP ribosylation, etc.) in genetic and alternate promoter regulation, as potential oncogene-inducing factors. Aberrant DNA methylation (hyper and hypo) of non-CpG island promoters, and genome wide hypo, can lead directly to cancer. Finally, interactions of various disease factors create large number of permutations, providing a diversity of genetic diseases. For instance, DNA methylation may cause histone deacetylation—and vice versa; and methyl binding proteins such as MeCP and MBD can deliver HDACs and HMTs that target, for instance, repressive lysines, deacetylate histones, or remodel chromatin.

CTCF–cohesin interaction with the DNA substrate could involve DNA loop(s) (influenced by insulators); or maybe spatially proximal chromatinized helixes (influenced by sister chromatid pairing). CTCF dimerization may provide the basis for MHC-II insulators "holding together" the stabilization of gene expression via levels of functional cohesin in the cell, and MHC-II insulators serving as nucleation points for the transcription functional complex—as well as cohesin regulation of transcription via stabilization of bound distal insulators with proximal promoter sequences. It is suggested that the cohesin ring structure is able to encircle the sister chromatids, providing structure and support. Additionally, cellular responses to RNAi silencing, and to the enzymatic processing of primary miRNA or dsRNA, represent additional mechanisms for dysregulation/disease of genes and cells when subject to improper "interference" of gene regulation and integrity. All such factors illustrate specific mechanisms for the dysregulation of genes and cells, and causations for disease.

Examples of genetic dysfunction related diseases involve single causative factors or various combinations of the following: uniparental disomy, deletions, imprint defects, point mutations, translocation mutations, alternate splicing and duplications. Some selected examples of specific diseases and their mechanisms include the following. Fragile X syndrome symptoms include retardation, large forehead and ears, with some neurodevelopmental phenotypes. Etiology is mapped to the "fragile" Xq27.3 chromosome; and FMR1 (encodes FMRP protein) "malexpression" causes unstable CGG repeat, mediated in cis (aberrant methylation, acetylation). Fragile X involves a primary mutation plus secondary epigenetic mutation; FMRP malexpression is also related to synaptic developmental issues. In Rett syndrome, a neurological disorder exhibiting multiple symptoms, the q27 arm of the X chromosome in the gene encoding MECP2 methyl binding protein is bound with methylated CpG dinucleotides, acting to repress transcription through recruitment of the Sin3A, SWI-SNF, and HDAC repressors by maintaining localized heterochromatin. The same epigenetic mechanisms discussed herein may also lead to a better understanding of complex diseases, in addition to those directly related to epigenetic abnormalities—and to therapies based on both histone modifications and DNA methylation and RNAi pathway mediation.

Epigenetics, as the totality of chromatin template alterations effecting transcription and silencing of genes from a common genome, extends our understanding of genomic bioinformatics and the biotechnology arena—along with our perspectives on causality, previously alleged "junk" DNA, and the fundamental considerations for use of cancer and stem cell lines. Epigenetics modifies the means by which the internal code of the genotype manifests its external product, the phenotype—resulting in diversity of form and function (morphology). [Ala the Central Dogma, DNA to RNA (transcription) to protein (translation) to traits]. Epigenetics describes the interactions of the genotype with the environment to produce the phenotype—extending the traditional Mendelian genetics of heredity with a much broader paradigm. The new paradigm includes a diversity of pathways linking phenotype to genotype. These include DNA–protein interactions, transcription–translation relationships based on temporal sequences of events and environments within and without the cell/nucleus, and molecular process stochastics. Epigenetic dialectic derives from the complexity of protein effectors modulated by histone modifications, and enzyme substrate specificity directing repression or activation of transcription; all bestowed with additional specificity through chromatin remodeling, HATs, DNMTs, HDACs, HDMs, DDMs, and DNA methylytransferases and demethylases—as well as even more specific transcription factors representing the genotype contribution to the epigenetic manifestation of the phenotype.

Significant changes in epigenetic coding occur during the development life cycle. Embryonic stem cells can reprogram the genome's epigenetic marks, providing diversity of gene regulation and expression (pluripotency), and potentially extending into successive generations; while adult stem cells can be used for more specialized purposes and medical therapies. Epigenetic programming involves intervention in the epigenetics of pluripotent stem cells. This can be accompanied by germ cell line production in culture for research or therapies—perhaps bringing us very close to the controversial, albeit illegal, practice of human cloning—where distinction might be needed for cells, tissues, embryos, and humans—and a legal delineation established for the question of where life begins? I propose it begins with consciousness. For example, Henrietta Lacks' cancer cells—while "immortal" are not conscious.

The idea of personalized embryonic stem cells addressing individual needs is certainly an attractive prospect. The key benefit to be derived from such research is obtained by identifying an abnormality and studying the associated epigenetics—rather than justifying the effort by the lives it may save; since the genetic diseases may present in only a very small percentage of the population, thereby obviating a prioritization based on saving lives.

But if by researching the effects, for instance, of retroviral insertion of genetic information into the nucleus, we discover a means to control or reverse a certain cancer—or to "heal" damaged brain tissue—then the focused research would indeed benefit many more lives than the expectation based on the specific research agenda at initiation. So when the biotechnologist indicates he or she is "working on a cure for cancer"—we cannot say with certainty this is not the case, despite the initially limited experimental domain. Biochemistry affords the merger of bioinformatics, genomics, epigenomics and proteomics—providing rich potential for rapid advancements—thereby further justifying the expansion of research in these areas. Undesired avoidance of apoptosis and growth prevention factors leading to cancer may provide positive benefit to proliferation and immortality of, for instance, a HeLa culture for epigenetic mapping experiments. Therefore, the benefits of directed research are not totally predictable.

Since mammals cannot reproduce via parthenogenesis, parent-specific epigenetic processes are needed for the genome during gametogenesis; i.e., differential expression of imprinting by the two parental genetic alleles, *unless*—and we know our exceptions are plenteous—the female gamete epigenotype can be manipulated, for example, to produce males that are exclusively of maternal origin. Perhaps that's "not the android we're looking for"—as natural processes of evolution and genetic conservation could lead to unanticipated results such as disease vulnerability or extinction. A totipotent mammal could indeed prove quite lacking in the raison d'être, or the claimed reason for existence, that provides the moral fabric upon which we base our ethical value system. Not until we've achieved a closed-form solution to the systematic relationships between genotype, phenotype, disease, and the environment should we propose optimization or global bioreprogramming routines. To this end— none of the DNA is garbage—as indirect and combinatorial processes may utilize any portions not currently translating to protein. We should not perturb the chromatin structure until we've solved and resolved the complete set of algorithms represented in the physical solution space. The epigenetic solution space involves putting garbage DNA to work in a productive manner, supervised in part by microRNA, environmental and epigenetic fabrics, and temporal sequencing of logical biosystem space, tempered by advantageous transfection of desirable information and knowledge.

We now examine transport and the nuclear pore complex (NPC). "Random" diffusion of the import complex is unidirectional, from the relatively high cytoplasmic concentration to the lower concentration in the nucleoplasm; and similarly the random diffusion across a concentration gradient occurs for importin transport out of the nucleoplasm to the cytoplasm, again from high to low concentration. Ran-GEF and Ran-GAP promote cargo dissociation in the nucleoplasm, and importin dissociation in the cytoplasm, respectively; serving to maintain the concentration gradients of cargo complex and Ran~GTP across the NPC—the macromolecular cylinder spanning the nuclear membranes.

Diffusion is thought to be enabled by hydrophobic FG-repeat structures called nucleoporins (FG-nucleoporins) lining the entire surface of the NPC basket.

The hydrophobic importin cargo complexes are able to diffuse through the basket, while hydrophilic proteins are slowed or precluded. I would suggest this may be due to the effect of elimination or minimization of hydrogen bonded networks transiently forming and breaking at the cargo–NPC interface, due to the bonding exclusion resulting from the hydrophilic chaperone complex; while at the same time the FG repeats transiently bond to NLSs. Unaccompanied hydrophilic proteins could form hydrogen bonds with surface areas not entirely "covered" by FGs, thereby slowing or preventing passive diffusion through the basket. This could be assisted via the dynamic reconfiguration of the Phe-Gly repeats in the Nups—and consistent with the consensus three-step mechanisms.

Or perhaps the Nups lining the basket exhibit some periodicity corresponding to the hydrophobic/hydrophilic sequence distributions along the cargo complex. This would be an interesting problem to model/simulate with computational molecular dynamics—and to then validate with an artificial nanotube Nup construct. Perhaps the clathrin and coat protein folds similar to those of the Nups would reveal the nature of the hydrophobic interactions occurring as the FG's bind to the cognate-shuttling proteins. These periodic interactions might serve as a transport mechanism that "moves the cargo complex along" as it diffuses through the NPC. The EM imaging that demonstrated the dynamic and flexible nature of the NPC itself is exciting—and the "nanoaquarium" developed at the University of Pennsylvania (Mauk, M., private conversation 2012) may provide a means to conduct in vivo dynamic images to further elucidate the NPC diffusion processes—as well as mRNA filtering effects.

Similar to the mechanism described for import, export from the nucleus to the cytoplasm involves shuttle proteins containing nuclear export and localization signal sequences (NES and NLS, respectively). First, a trimolecular complex is formed in the nucleus, consisting of Ran~GTP, exportin 1 (nuclear-export receptor), and the cargo to be exported. Perhaps in a manner similar to the hydrophobic effect described for import above, the trimolecular complex diffuses (again from high to low concentration gradient of cargo complex from high to low) through the NPC to the cytoplasm. There, the cargo dissociates from the complex via Ran-GAP hydrolization of the Ran~GTP, and release of the complex constituents. One notable difference between the import and export processes is the Ran~GTP presence in the cargo complex for export, but not for import.

The hydrophobic karyopherin family of importins and exportins is highly conserved; and some of the cognate NES/NLS proteins are seen to function as both importins and exportins. The shuttling mechanisms for alternate cargos resemble the processes described—and most are Ran-dependent. Thus the NPC serves the role of transporting cargo back and forth across the nuclear envelope. As we see in the next question, the NPC also delivers genetic programming to the cytoplasm for protein synthesis.

In addition, epigenetic controls associated with the NPC include chromatin boundary delineation, modulation of transition regions of chromatin density distributions associated with varied transcription and/or mRNA genesis and repair "stations." It is hypothesized that the NPC serves as a staging area for congregation of active genes and nucleation sites for biogenesis. Again, this staging area is replete with chromatin modeling complexes—also recruited to the "loading dock," to be activated, modulated, regulated, and "shipped out." Add to this, chromatin maintenance and repair, including telomere functions, and finally chromatin segregation during mitosis— and we see the vast array of functionality at the NPC, and the specifics of the gene expression portion of this activity.

REFERENCE

Lodish, H., A. Berk, C.A. Kaiser, et al. 2008. *Molecular cell biology*, 6th edition. New York: W.H. Freeman and Company.

Appendix 4
Rhodopsin and Signal Transduction

Some time ago, Mentzer (previously unpublished) envisioned eye drops with ligand effectors binding to protein disruptions in the rhodopsin chain, as an immediate indicator of mTBI. Clinical pupillary measurements could be enhanced with a colorimetric quantification based on changes related to mTBI. Correlates with eye tracking devices for mTBI assessment would benefit as well.

Rhodopsin is a light-activated G protein-coupled receptor (GPCR) in rod cells of the eye—specifically located in flat discs in the outer portion of the rod cells. The rhodopsin GPCR is covalently bound to 11-cis-retinal, which is a visual pigment that responds to the "visible" portion of the electromagnetic spectrum, and is completely surrounded by the seven membrane spanning regions of the GPCR. In the case of rhodopsin the coupled trimeric G protein is transducin (Gt), with the Ga class subunit Gat (Lodish et al., 2008).

The 11-cis-retinal lysine side chain couples to opsin in the rhodopsin, which becomes activated through absorption of a photon of light, concurrent with the 11-cis-retinal isomerization to the all-trans-retinal moiety, now coupled to the activated opsin, in what we call meta-rhodopsin II. Note that the 11-cis-retinal couples to the opsin's lysine side chain, and all-trans retinal is restored to 11-cis-retinal.

This meta-stable intermediate activates the Gt protein's Gat subunit. After several seconds, the trans moiety dissociates from opsin and converts back to the cis moiety for the subsequent binding to inactivated opsin. Analogous to the conformational change associated with ligand binding in other GPCRs, there is a conformational change in the opsin upon photon activation. Light absorption causes nonselective ion channels (for Na^+, Ca^{2+}, K^+) in the rod membrane to close, polarizing the resting potential of the membrane to a higher inside negativity, which is transmitted to the brain and perceived as light (Lodish et al., 2008).

Light absorption induced closing of the nonselective ion channels in the plasma membrane of the rod cells involves the cGMP second messenger, which keeps the channels open in the absence of light. Light-activated opsin, bound to Gat, mediates GDP>GTP; and activates cGMP phosphodiesterase (PDE), which converts cGMP to GMP. The drop in cytosolic cGMP causes the ion channels to close, leading to the electrical polarization transient across the rod membrane that is detected in the brain as light. So, the PDE is the effector protein, and the cGMP is the secondary messenger (Lodish et al., 2008).

Having outlined the variant of the GPCR represented by rhodopsin and conversion of optical photons to creation of visual images in the brain, we are equipped to properly assess the work of Smith (2010). Smith reviews rhodopsin as a special case of the GPCR, analyzes signal transduction from a 3D structural perspective, provides a structural explanation for the Ga class genetically conserved amino acids and structural motifs, and more specifically, conserved residues in visual receptors, leading to convergence with the more general models for the Ga class of GPCRs. Smith focuses especially on the activation mechanism in rhodopsin, remarkable in many respects—including a photoreaction quantum yield of 0.67 (better than our best engineered photovoltaic solar cells) and a photoisomerization process that occurs in 200 femtoseconds).

The most loosely packed of the beta-sheet folds in the seven transmembrane helices (H5 and H6) are shown to undergo the largest reconfiguration (displacement) upon activation of the 11-cis-retinal. Hydrogen bonded networks are of great importance in the maintenance of a high dissociation constant of the protonated Schiff's base linking the 11-cis-retinal to the receptor protein; as well as importance in stabilization of the extracellular loops; and finally to their rearrangement upon isomerization/activation. Indeed, I would note the quantum yield for 11-cis-retinal in solution is only 0.3, and accompanied by multiple isomerizations. Hydrogen bonding in vivo certainly contributes to stabilization and reconfigurations of the four molecular switch "microdomains" observed in rhodopsin (Smith, 2010).

The absorption band shifts observed during the thermal relaxation of the photoreaction intermediates, further defined as a "series of distinct, spectrally defined intermediates," provides numerous intriguing insights into overall signal transduction—namely, the relation of wavelength shift to protonation of the Schiff's base, wavelength/conformation relationships and controls, temporal characterization of reaction intermediates, motion of H5 and H6 within the protein binding pocket, and ultimately, the ability to ascertain the function of conserved residues to validate the generalized GPCR models proposed for retinal. This leads to improved understanding of disease states with respect to mutations of these conserved residues—along with avenues for therapeutic drug targets. It may also lead to insights regarding the GTP > GDP exchange mechanism, which is not fully understood in terms of signal transduction mechanism (Smith, 2010).

Smith illustrates, in a series of motif, domain, conformation -driven reaction sequences, conserved functionality identification, and water-mediated hydrogen bonding stabilizations; how rhodopsin evolved as a photoreceptor with remarkable dynamic range and sensitivity. He points the way for further investigation into the nature of the generalized Ga class of trimeric G proteins in GPCRs as the basis for improved understanding of the unique specifics of the rhodopsin receptor. Rather than signal transduction via movement of an electrical or chemical potential through a reaction mechanism circuit, signal transduction in rhodopsin is seen as a sequence of consecutive "information" transfers involving modulation of binding potentials, metastable intermediates, conformational signals and effectors, and direct control activation via light photons rather than ligand binding.

Further to the specifics of the second messenger phototransduction cascade in rhodopsin photopigment, some additional detail of what happens in the activation process in the disk membrane goes like this:

~hv >> 11-cis to all-trans retinal isomer >> transducin activation >> phosphodiesterase (PDE) activation >> hydrolysis of cGMP, reducing cGMP concentration available to bind to channels >> closing of channels in outer membrane >> producing differential cation transients across the membrane >> producing transient hyperpolarization transients.

And deactivation proceeds like this:

rhodopsin kinase >> phosphorylation of active rhodopsin >> permitting arrestin to bind to rhodopsin >> blocking transducin activation by activated rhodopsin >> ending the phototransduction cascade

Via the retinoid cycle, after photoisomerization, retinal is restored back to the all-trans form like this:

all trans retinal >> converts to all trans retinol >> which is transported by interphotoreceptor retinoid binding protein (IRBP) chaperone into pigment epithelium >> where it gets transformed to 11-cis retinal >> and then chaperoned again by IRBP to the outer membrane segment >> and combined with opsin.

Additional signaling pathways include the light adaptation process. This results in greater gray-scale sensitivity over a wider dynamic range, analogous to photomultipliers rather than the limited performance of, for instance, a conduction mode pn junction diode—analogizing to the solid-state detector world, in which we've not achieved comparable success.

Further to the concept of nuclear signaling—we can separate the mammalian code—of approximately 25,000 genes, translating to more than 200 different cell types—into two groups: germ cells and somatic cells. Germ cells have the ability to divide and produce all the other cell types, and somatic cells represent the specific engines of life. The epigenome is modulated by activation of eight major classes of cell surface receptors, translating signaling molecules into cellular transcription response.

After the cell cycle progresses to the point where somatic differentiation results in cells progressing beyond the germ and stem cell phase, the transcription factors are influenced by two categories of signals: First is the set of external signals we've been discussing, to include cytokines, growth factors, hormones, etc., along with stress type responses (Lodish et al., 2008); and second is a set of intrinsic factors to include transcription itself, replication of DNA, and chromosome modifications and segregation. Polycomb and trithorax (PcG and trxG) are two regulators of differentiation and cell specification in the eukaryotes (Allis et al., 2009).

Histone modifications result in repression or alteration of proper functioning of the genome (chromatin deregulation) and, along with our extrinsic factor regulation, may collectively supply the approach to a personalized therapy for a wide range of maladies. In fact, the epigenomic code, when established, may provide the basis for determination of which kinase cascades are malfunctional and require "adjustments".

Transprocess signaling between the extrinsic and intrinsic factors discussed provides an additional regulatory signaling mechanism. Therefore—just as we've looked at how protein domain functionality assists with cascades of kinase reactions—the epigenetic condition determines which portions of the genetic sequence are available to be transcribed.

Recent progress in the field of optogenetics (Farmer, 2020; Zhang et al., 2011) provides the means to activate or deactivate transmembrane rhodopsins to manipulate neuronal circuits at the cellular level. This provides better understanding of neuronal circuits and brain synapses.

REFERENCES

Allis, C.D., T. Jenuwein, D. Reinberg, and M. Caparros, eds. 2009. *Epigenetics.* Cold Spring Harbor Press, New York. pp. 43–50.

Farmer, D. 2020. Opsins travel to the brain's hidden places. A rising number of light-sensitive proteins and optogenetics are enabling precise imaging of brain cells, as well as the potential to adapt functioning in neuronal networks. *BioPhotonics.* Pittsfield, MA: Lauren Publishing. https://www.photonics.com/Articles/ Opsins_Travel_to_the_Brains_Hidden_Places/a65549

Lodish, H., A. Berk, C.A. Kaiser, et al. 2008. *Molecular cell biology,* 6th edition. New York: W.H. Freeman and Company.

Smith, S. 2010. Structure and activation of the visual pigment rhodopsin. *Annu Rev Biophys.* 39:309–328.

Zhang, F., J. Vierock, O. Yizhar, et al. 2011. The microbial opsin family of optogenetic tools. *Cell.* 147(7):1446–1457.

Index

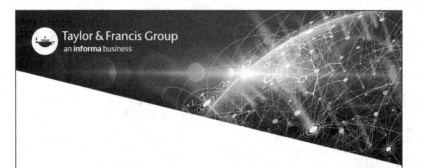

Printed in the United States
By Bookmasters